Lecture Notes in Artificial Intelligence 2148

Subseries of Lecture Notes in Computer Science
Edited by J. G. Carbonell and J. Siekmann

Lecture Notes in Computer Science

Edited by G. Goos, J. Hartmanis, and J. van Leeuwen

W0049878

Springer
Berlin
Heidelberg
New York
Barcelona
Hong Kong
London
Milan
Paris
Tokyo

Alexander Nareyek (Ed.)

Local Search for Planning and Scheduling

ECAI 2000 Workshop
Berlin, Germany, August 21, 2000
Revised Papers

Springer

Series Editors

Jaime G. Carbonell, Carnegie Mellon University, Pittsburgh, PA, USA
Jörg Siekmann, University of Saarland, Saarbrücken, Germany

Volume Editor

Alexander Nareyek
GMD FIRST
Kekuléstr. 7, 12489 Berlin, Germany
E-mail: alex@ai-center.com

Cataloging-in-Publication Data applied for

Die Deutsche Bibliothek - CIP-Einheitsaufnahme

Local search for planning and scheduling : revised papers / ECAI 2000
Workshop, Berlin, Germany, August 21, 2000. Alexander Nareyek (ed.). -
Berlin ; Heidelberg ; New York ; Barcelona ; Hong Kong ; London ; Milan ;
Paris ; Tokyo : Springer, 2001
    (Lecture notes in computer science ; Vol. 2148 : Lecture notes in
    artificial intelligence)

CR Subject Classification (1998): I.2.8, F.2.2, G.1.6, G.2.2

ISBN 3-540-42898-4 Springer-Verlag Berlin Heidelberg New York

This work is subject to copyright. All rights are reserved, whether the whole or part of the material is
concerned, specifically the rights of translation, reprinting, re-use of illustrations, recitation, broadcasting,
reproduction on microfilms or in any other way, and storage in data banks. Duplication of this publication
or parts thereof is permitted only under the provisions of the German Copyright Law of September 9, 1965,
in its current version, and permission for use must always be obtained from Springer-Verlag. Violations are
liable for prosecution under the German Copyright Law.

Springer-Verlag Berlin Heidelberg New York
a member of BertelsmannSpringer Science+Business Media GmbH

http://www.springer.de

© Springer-Verlag Berlin Heidelberg 2001

Typesetting: Camera-ready by author, data conversion by PTP-Berlin, Stefan Sossna
Printed on acid-free paper   SPIN: 10845478   06/3142   5 4 3 2 1 0

Preface

With the increasing deployment of planning and scheduling systems, developers often have to deal with very large search spaces, real-time performance demands, and dynamic environments. Complete refinement methods do not scale well, making local search methods the only practical alternative. A dynamic environment also promotes the application of local search, the search heuristics not normally being affected by modifications of the search space. Furthermore, local search is well suited for anytime requirements because the optimization goal is improved iteratively. Such advantages are offset by the incompleteness of most local search methods, which makes it impossible to prove the inconsistency or optimality of the solutions generated. Popular local search approaches include evolutionary algorithms, simulated annealing, tabu search, min-conflicts, GSAT, and Walksat. The first article in this book – an invited contribution by Stefan Voß – gives an overview of these methods.

The book is based on the contributions to the Workshop on Local Search for Planning & Scheduling, held on August 21, 2000 at the 14th European Conference on Artificial Intelligence (ECAI 2000) in Berlin, Germany. The workshop brought together researchers from the planning and scheduling communities to explore these topics with respect to local search procedures. After the workshop, a second review process resulted in the contributions to the present volume.

Voß's overview is followed by two articles, by Hamiez and Hao and Gerevini and Serina, on specific "classical" combinatorial search problems. The article by Hamiez and Hao addresses the problem of sports-league scheduling, presenting results achieved by a tabu search method based on a neighborhood of value swaps. Gerevini and Serina's article addresses the topic that dominates the rest of the book: action planning. It builds on their previous work on local search on planning graphs, presenting a new search guidance heuristic with dynamic parameter tuning.

The next set of articles deal with planning systems that are able to incorporate resource reasoning. The first article, of which I am the author, makes it clear why conventional planning systems cannot properly handle planning with resources and gives an overview of the constraint-based EXCALIBUR agent's planning system, which does not have these restrictions. The next three articles are about NASA JPL's ASPEN/CASPER system. The first one – by Chien, Knight, and Rabideau – focuses on the replanning capabilities of local search methods, presenting two empirical studies in which a continuous planning process clearly outperforms a restart strategy. The next article, by Engelhardt and Chien, shows how learning can be used to speed up the search for a plan. The goal is to find a set of search heuristics that guide the search as well as possible. The last article in this block – by Knight, Rabideau, and Chien – proposes and demonstrates, a technique for aggregating single search moves so that distant states can be reached more easily.

The last three articles in this book address topics that are not directly related to local search, but the described methods make very local decisions during the search. Refanidis and Vlahavas describe extensions to the GRT planner, e.g., a hill-climbing strategy for action selection. The extensions result in much better performance than with the original GRT planner. The second article – by Onaindia, Sebastia, and Marzal – presents a planning algorithm that successively refines a start graph by different phases, e.g., a phase to guarantee completeness. In the last article, Hiraishi and Mizoguchi present a search method for constructing a route map. Constraints with respect to memory and time can be incorporated into the search process.

I wish to express my gratitude to the members of the program committee, who acted as reviewers for the workshop and this volume. I would also like to thank all those who helped to make this workshop a success – including, of course, the participants and the authors of papers in this volume.

June 2001   Alexander Nareyek
                                                                  Workshop Chair

Program Committee

Emile H. L. Aarts   Philips Research
José Luis Ambite   Univ. of Southern California
Blai Bonet   University of California
Ronen I. Brafman   Ben-Gurion University
Steve Chien   NASA JPL
Andrew Davenport   IBM T. J. Watson
Alfonso Gerevini   Università di Brescia
Holger H. Hoos   Univ. of British Columbia
Alexander Nareyek   GMD FIRST
Angelo Oddi   IP-CNR
María C. Riff   Univ. Téc. Fed. Santa María
Bart Selman   Cornell University
Edward Tsang   University of Essex

Table of Contents

Meta-heuristics: The State of the Art

Stefan Voß

Technische Universität Braunschweig
Institut für Wirtschaftswissenschaften
Abt-Jerusalem-Straße 7
D-38106 Braunschweig, Germany
stefan.voss@tu-bs.de

Abstract. Meta-heuristics support managers in decision-making with robust tools that provide high-quality solutions to important applications in business, engineering, economics and science in reasonable time horizons. In this paper we give some insight into the state of the art of meta-heuristics. This primarily focuses on the significant progress which general frames within the meta-heuristics field have implied for solving combinatorial optimization problems, mainly those for planning and scheduling.

1 Introduction

Many decision problems in business and economics, notably including those in manufacturing, location, routing, and scheduling may be formulated as optimization problems. Typically these problems are too difficult to be solved exactly within a reasonable amount of time and heuristics become the methods of choice. In cases where simply obtaining a feasible solution is not satisfactory, but where the quality of solution is critical, it becomes important to investigate efficient procedures to obtain the best possible solutions within time limits deemed practical.

Due to the complexity of many of these optimization problems, particularly those of large sizes encountered in most practical settings, exact algorithms often perform very poorly (in some cases taking days or more to find moderately decent, let alone optimal, solutions even to fairly small instances). As a result, heuristic algorithms are conspicuously preferable in practical applications.

Among the most studied heuristics are those based on applying some sort of greediness or applying priority based procedures including, e.g., insertion and dispatching rules. As an extension of these, a large number of local search approaches has been developed to improve given feasible solutions. The main drawback of these approaches, their inability to continue the search upon becoming trapped in a local optimum, leads to consideration of techniques for guiding known heuristics to overcome local optimality. Following this theme, one may investigate the application of intelligent search methods like the tabu search meta-heuristic for solving optimization problems.

In this paper we sketch the state of the art of meta-heuristics mainly from an operations research perspective, knowing that probably any attempt to be fully

A. Nareyek (Ed.): Local Search for Planning and Scheduling, LNAI 2148, pp. 1–23, 2001.
© Springer-Verlag Berlin Heidelberg 2001

comprehensive in this respect within a single paper has to fail. Moreover, we try to provide some insights and highlight most of the areas of interest with the primary focus on the significant progress which general frames within the meta-heuristics field have implied for solving combinatorial optimization problems and provide important references.

In Section 2 we present an introduction to *heuristics* and *meta-heuristics* and provide necessary preliminaries. In Section 3 we describe the ingredients and basic concepts of various strategies like *tabu search, simulated annealing* and *genetic algorithms*. This is based on a simplified view of a possible inheritance tree for heuristic search methods, illustrating the relationships between some of the most important methods discussed below, as shown in Figure 1.

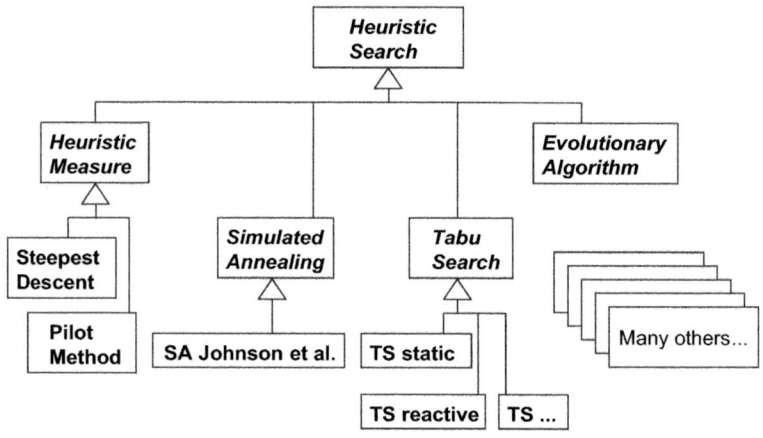

Fig. 1. Simplified meta-heuristics inheritance tree.

This is followed by some recent advances including the important incorporation of exact methods into intelligent search. The main focus of this section (and the overall paper) is to show that general frames such as *adaptive memory programming* exist that may subsume various approaches within the meta-heuristics field. Finally, some conclusions and ideas for future research are given.

2   Heuristics: Patient Rules of Thumb and Beyond

The basic concept of heuristic search as an aid to problem solving was first introduced by [83]. A *heuristic*[1] is a technique (consisting of a rule or a set of rules) which seeks (and hopefully finds) *good* solutions at a reasonable computational cost. A heuristic is *approximate* in the sense that it provides (hopefully) a good solution for relatively little effort, but it does not guarantee optimality.

---

[1] In Greek we have *heuriskein*: to find, to discover.

Reviewing the literature, there is a great variety of definitions that more or less conform with our introductory remark (see, e.g., [73,79,87,118]). Moreover, the usual distinction refers to finding initial feasible solutions and improving them. Arguing why heuristics are useful need not be considered here as obvious reasons are widely discussed throughout the literature (see, e.g., [30] for complexity issues).

2.1   Intuition (Starting Greedily)

Heuristics provide simple means of indicating which among several alternatives seems to be the best. And basically they are based on intuition. That is, "heuristics are criteria, methods, or principles for deciding which among several alternative courses of action promises to be the most effective in order to achieve some goal. They represent compromises between two requirements: the need to make such criteria simple and, at the same time, the desire to see them discriminate correctly between good and bad choices. A heuristic may be a *rule of thumb* that is used to guide one's action." [79]

Greedy heuristics are simple heuristics available for any kind of combinatorial optimization problem. They are iterative and a good characterization is their *myopic* behavior. A greedy heuristic starts with a given feasible or infeasible solution. In each iteration there is a number of alternative choices (*moves*) that can be made to transform the solution. From these alternatives which consist in fixing (or changing) one or more variables, a *greedy choice* is made, i.e., the best alternative according to a given evaluation measure is chosen until no such transformations are possible any longer.

Usually, a greedy construction heuristic starts with an incomplete solution and completes it stepwise. Savings and dual algorithms follow the same iterative scheme: Dual heuristics change an infeasible low cost solution until reaching feasibility, savings algorithms start with a high cost solution and realize the highest savings as long as possible. Moreover, in all three cases, once an element is chosen this decision is not reversed throughout the algorithm, it is kept.

As each alternative has to be measured, in general we may define some sort of *heuristic measure* (see, e.g., [20]) which is iteratively followed until a complete solution is build. Usually this heuristic measure is performed in a greedy fashion.

2.2   Local Search

The basic principle of local search is that solutions are successively changed by performing moves which alter solutions *locally*. Valid transformations are defined by neighborhoods which give for a solution all neighboring solutions that can be reached by one move.[2] For an excellent general survey on local search see the collection of [1] and the references in [2]. A simple template is provided by [104].

---

[2] Formally, we consider an instance of a combinatorial optimization problem with a solution space S of feasible (or even infeasible) solutions. To maintain information about solutions, there may be one or more solution information functions I on S, which are termed exact, if I is injective, and approximate otherwise. With this

Moves must be evaluated by some *heuristic measure* to guide the search. Often one uses the implied change of the objective function value, which may provide reasonable information about the (local) advantage of moves. Following a greedy strategy, *steepest descent* (SD) corresponds to selecting and performing in each iteration the best move until the search stops at a local optimum. Obviously, savings algorithms (see Section 2.1) correspond to SD.[3]

As the solution quality of local optima may be unsatisfactory, we need mechanisms which guide the search to overcome local optimality. A simple strategy called iterated local search is to iterate/restart the local search process after a local optimum has been obtained, which requires some perturbation scheme to generate a new initial solution (e.g., performing some random moves). Of course, more structured ways to overcome local optimality might be advantageous.

Starting in the 1970s [65], a variable way of handling neighborhoods is still a topic within local search. Consider an arbitrary neighborhood structure N, which defines for any solution s a set of neighbor solutions $N_1(s)$ as a neighborhood of depth $d = 1$. In a straightforward way, a neighborhood $N_{d+1}(s)$ of depth $d + 1$ is defined as the set $N_d(s) \cup \{s' | \exists s'' \in N_d(s) : s' \in N_1(s'')\}$. In general, a large d might be unreasonable, as the neighborhood size may grow exponentially. However, depths of two or three may be appropriate. Furthermore, temporarily increasing the neighborhood depth has been found to be a reasonable mechanism to overcome *basins of attraction*, e.g., when a large number of neighbors with equal quality exist.

Recently, *large scale neighborhoods* have become an interesting research topic (see, e.g., [3] for a survey), especially when efficient ways are at hand for exploring them. Related research can also be found under various names; see, e.g., [84] for the idea of dynasearch and [81] for so-called ejection chains.

To give an example, dynasearch uses dynamic programming to search an exponential size neighborhood in polynomial time. Computational results for some scheduling problems show that, for restarts close to previous local minima, dynasearch can be significantly effective. As a consequence, one may try to use dynasearch with an iterated local search framework, where each descent is started a few random moves away from previous local minima.

2.3   Meta-heuristics

The formal definition of meta-heuristics is based on a variety of definitions from different authors based on [32]. Basically, a meta-heuristic is a top-level strategy that guides an underlying heuristic solving a given problem. Following Glover

---

information, one may store a search history (trajectory). For each S there are one or more neighborhood structures N that define for each solution $s \in S$ an ordered set of neighbors $N(s) = \{n_1(s), \ldots, n_{|N(s)|}(s)\}$. To each neighbor $n(s) \in N(s)$ corresponds a move that captures the transitional information from s to $n(s)$.

[3] In the sense that a transition or move may be performed according to a neighborhood search where S need not consist of feasible solutions only (i.e., we allow for some *strategic oscillation* between feasibility and infeasibility), we may even overcome the distinction between heuristics and improvement or local search procedures.

it "refers to a master strategy that guides and modifies other heuristics to produce solutions beyond those that are normally generated in a quest for local
optimality." [37] In that sense we distinguish between a *guiding process* and an
application process. The guiding process decides upon possible (local) moves and
forwards its decision to the application process which then executes the chosen
move. In addition, it provides information for the guiding process (depending
on the requirements of the respective meta-heuristic) like the recomputed set of
possible moves.

According to [37] "meta-heuristics in their modern forms are based on a
variety of interpretations of what constitutes *intelligent search,*" where the term
intelligent search has been made prominent by Pearl [79] (regarding heuristics
in an artificial intelligence context) and [108] (regarding an operations research
context). In that sense we may also consider the following definition: "A meta-
heuristic is an iterative generation process which guides a subordinate heuristic
by combining intelligently different concepts for exploring and exploiting the
search spaces using learning strategies to structure information in order to find
efficiently near-optimal solutions." [77]

To summarize, the following definition seems to be most appropriate: *"A
meta-heuristic is an iterative master process that guides and modifies the operations of subordinate heuristics to efficiently produce high-quality solutions.
It may manipulate a complete (or incomplete) single solution or a collection
of solutions at each iteration. The subordinate heuristics may be high (or low)
level procedures, or a simple local search, or just a construction method. The
family of meta-heuristics includes, but is not limited to, adaptive memory procedures, tabu search, ant systems, greedy randomized adaptive search, variable
neighborhood search, evolutionary methods, genetic algorithms, scatter search,
neural networks, simulated annealing, and their hybrids."* ([111], p. ix)

Of course the term meta-heuristics has also been used in different other con-
texts (see, e.g., [64]) or even misused or misunderstood throughout the literature.
Examples are found in various directions. One of the misleading understandings
is as follows (see, e.g., [59]): Given an instance of a combinatorial (optimization)
problem and m different heuristics (or algorithms) $H_1, H_2, \ldots, H_m$ the heuristic
H_{best} will output the best among the m outputs of heuristics $H_1, H_2, \ldots, H_m$.
H_{best} as the resulting heuristic is sometimes referred to as meta-heuristic.

3   Meta-heuristics Concepts

In this section we summarize the basic concepts of the most important meta-
heuristics. Here we shall see that adaptive processes originating from different
settings such as psychology ("learning"), biology ("evolution"), physics ("annealing"), and neurology ("nerve impulses") have served as a starting point.

Applications of meta-heuristics are almost uncountable and appear in various
journals (including the *Journal of Heuristics* and the *INFORMS Journal on
Computing*), books, and technical reports every day. A helpful source for a subset
of successful applications may be special issues of journals or compilations such

as [107,82,85,63,77,86,36,111], just to mention some. A useful meta-heuristics survey up to the mid 1990s is compiled by [78].

3.1   Simple Local Search Based Meta-heuristics

To improve the efficiency of greedy heuristics, one may apply some generic strategies that may be used alone or in combination with each other, namely changing the definition of alternative choices, look ahead evaluation, candidate lists, and randomized selection criteria bound up with *repetition*, as well as combinations with local search or other methods.

Greedy Randomized Adaptive Search (GRASP): If we omit a greedy choice criterion for a random strategy we can run the algorithm several times and obtain a large number of different solutions. However, a pure random choice may perform very poor on average. Thus a combination of best and random choice seems to be appropriate: We may define a *candidate list* consisting of a number of best alternatives. Out of this list one alternative is chosen randomly. The length of the candidate list is given either as an absolute value, a percentage of all feasible alternatives or implicitly by defining an allowed quality gap (to the best alternative), which also may be an absolute value or a percentage.

Replicating a search procedure to determine a local optimum multiple times with different starting points has been acronymed as GRASP and investigated with respect to different applications, e.g., by [22]. The different initial solutions or starting points are found by a greedy procedure incorporating a probabilistic component. That is, given a candidate list to choose from, GRASP randomly chooses one of the best candidates from this list in a greedy manner, but not necessarily the best possible choice.

The underlying principle is to investigate many good starting points through the greedy procedure and thereby to increase the possibility of finding a good local optimum on at least one replication. The method is said to be adaptive as the greedy function takes into account previous decisions when performing the next choice. It should be noted that GRASP goes back to older approaches [43].

The Pilot Method: Building on a simple greedy algorithm such as a construction heuristic the *pilot method* [21,20] is another meta-heuristic. It builds primarily on the idea to *look ahead* for each possible local choice (by computing a so-called "pilot" solution), memorizing the best result, and performing the according move. One may apply this strategy by successively performing a cheapest insertion heuristic for all possible local steps (i.e., starting with all incomplete solutions resulting from adding some not yet included element at some position to the current incomplete solution). The look ahead mechanism of the pilot method is related to increased neighborhood depths as the pilot method exploits the evaluation of neighbors at larger depths to guide the neighbor selection at depth one.

Usually, it is reasonable to restrict the pilot process to a given *evaluation depth*. That is, the pilot method is performed up to an incomplete solution (e.g., partial assignment) based on this evaluation depth and then completed by continuing with a conventional cheapest insertion heuristic. (It should be noted that similar ideas have been investigated under the acronym *rollout method* [10].) Some applications in continuous flow-shop scheduling can be found in [23], and some considerations regarding stochastic scheduling problems are given in [9].

Variable Neighborhood Search (VNS): For an excellent treatment of various aspects of VNS see [42]. They examine the idea of changing the neighborhood during the search in a systematic way. VNS explores increasingly distant neighborhoods of the current incumbent solution, and jumps from this solution to a new one iff an improvement has been made. In this way often favorable characteristics of incumbent solutions, e.g., that many variables are already at their optimal value, will be kept and used to obtain promising neighboring solutions.

Moreover, a local search routine is applied repeatedly to get from these neighboring solutions to local optima. This routine may also use several neighborhoods. Therefore, to construct different neighborhood structures and to perform a systematic search, one needs to have a way for finding the distance between any two solutions, i.e., one needs to supply the solution space with some metric (or quasi-metric) and then induce neighborhoods from it.

3.2   Simulated Annealing

Simulated annealing (SA) extends basic local search by allowing moves to inferior solutions [62,18]. The basic algorithm of SA may be described as follows: Successively, a candidate move is randomly selected; this move is accepted if it leads to a solution with a better objective function value than the current solution, otherwise the move is accepted with a probability that depends on the deterioration Δ of the objective function value. The probability of acceptance is computed as $e^{-\Delta/T}$, using a temperature T as control parameter.

Various authors describe a robust concretization of this general SA procedure. Following [57], the value of T is initially high, which allows many inferior moves to be accepted, and is gradually reduced through multiplication by a parameter *coolingFactor* according to a geometric cooling schedule. At each temperature *sizeFactor* $\times |N|$ move candidates are tested ($|N|$ denotes the current neighborhood size and *sizeFactor* an appropriate parameter). The starting temperature is determined as follows: Given a parameter *initialAcceptanceFraction* and based on an abbreviated trial run, the starting temperature is set so that the fraction of accepted moves is approximately *initialAcceptanceFraction*. A further parameter, *frozenAcceptanceFraction* is used to decide whether the annealing process is *frozen* and should be terminated. Every time a temperature is completed with less than *frozenAcceptanceFraction* of the candidate moves accepted, a counter is increased by one. The procedure is terminated when no new best solution is found for a certain value of this counter.

For a different concretization of SA see [54]. An interesting variant of SA is to strategically reheat the process, i.e., to perform a non-monotonic acceptance function. Successful applications are provided, e.g., in [76].

Threshold accepting [19] is a modification (or simplification) of SA with the essential difference between the two methods being the acceptance rules. Threshold accepting accepts every move that leads to a new solution which is 'not much worse' (i.e., deteriorates not more than a certain threshold which reduces with a temperature) than the older one.

3.3   Tabu Search

The basic paradigm of *tabu search* (TS) is to use information about the search history to guide local search approaches to overcome local optimality (see [37] for a survey on TS). In general, this is done by a dynamic transformation of the local neighborhood. Based on some sort of memory certain moves may be forbidden, we say they are set tabu. As for SA, the search may lead to performing deteriorating moves when no improving moves exist or all improving moves of the current neighborhood are set tabu. At each iteration a best admissible neighbor may be selected. A neighbor, respectively a corresponding move, is called admissible, if it is not tabu or if an aspiration criterion is fulfilled.

Below, we briefly describe various TS methods that differ especially in the way in which the tabu criteria are defined, taking into consideration the information about the search history (performed moves, traversed solutions). An aspiration criterion may override a possibly unreasonable tabu status of a move. For example, a move that leads to a neighbor with a better objective function value than encountered so far should be considered as admissible.

The most commonly used TS method is based on a *recency-based* memory that stores moves, more exactly move attributes, of the recent past (*static TS*, tabu navigation method). The basic idea of such approaches is to prohibit an appropriately defined inversion of performed moves for a given period. For example, one may store the solution attributes that have been created by a performed move in a tabu list. To obtain the current tabu status of a move to a neighbor, one may check whether (or how many of) the solution attributes that would be destroyed by this move are contained in the tabu list.

Strict TS embodies the idea of preventing cycling to formerly traversed solutions. The goal is to provide necessity and sufficiency with respect to the idea of not revisiting any solution. Accordingly, a move is classified as tabu iff it leads to a neighbor that has already been visited during the previous part of the search. There are two primary mechanisms to accomplish the tabu criterion: First, we may exploit logical interdependencies between the sequence of moves performed throughout the search process, as realized by the reverse elimination method and the cancellation sequence method (cf., e.g., [33,109,110]). Second, we may store information about all solutions visited so far. This may be carried out either exactly or, for reasons of efficiency, approximately (e.g., by using hash codes).

Reactive TS aims at the automatic adaptation of the tabu list length of static TS [8]. The idea is to increase the tabu list length when the tabu memory indicates that the search is revisiting formerly traversed solutions. A possible specification can be described as follows: Starting with a tabu list length l of 1 it is increased every time a solution has been repeated. If there has been no repetition for some iterations, we decrease it appropriately. To accomplish the detection of a repetition of a solution, one may apply a trajectory based memory using hash codes as for strict TS.

For reactive TS [8], it is appropriate to include means for diversifying moves whenever the tabu memory indicates that we are trapped in a certain region of the search space. As a trigger mechanism one may use, e.g., the combination of at least two solutions each having been traversed three times. A very simple escape strategy is to perform randomly a number of moves (depending on the average of the number of iterations between solution repetitions);[4] more advanced strategies may take into account some long-term memory information (like the frequencies of appearance of specific solution attributes in the search history).

Of course there is a great variety of additional ingredients that may make TS work successful as, e.g., restricting the number of neighbor solutions to be evaluated (using candidate list strategies).

The area of planning and scheduling encompasses a bewilderingly large variety of problems. The remarkable success of TS in the 1980s has nowhere been marked more than in the area of machine scheduling and sequencing with an evermore growing importance in real-world manufacturing (for a recent survey on TS in manufacturing see, e.g., [4]).

3.4   Evolutionary Algorithms

Evolutionary algorithms comprise a great variety of different concepts and paradigms including genetic algorithms (see, e.g., [48,38]), evolutionary strategies (see, e.g., [47,93]), evolutionary programs [27], scatter search (see, e.g., [31,34]), and memetic algorithms [71]. For surveys and references on evolutionary algorithms see also [28,6,72,67]. We restrict ourselves to very few comments.

Genetic algorithms are a class of adaptive search procedures based on the principles derived from the dynamics of natural population genetics. One of the most crucial ideas for a successful implementation of a genetic algorithm (GA) is the representation of an underlying problem by a suitable scheme. A GA starts (for example) with a randomly created initial population of artificial creatures (strings), found, for example, by flipping a 'fair' coin. These strings in whole and in part are the base set for all subsequent populations. They are copied and information is exchanged between the strings in order to find new solutions of the underlying problem. The mechanisms of a simple GA essentially consist of copying strings and exchanging partial strings. A simple GA requires three operators which are named according to the corresponding biological mechanisms: reproduction, crossover, and mutation. Performing an operator may depend on a *fitness function* or its value (fitness), respectively. This function defines a means of measurement for the profit or the quality of the coded solution for the underlying problem and may depend on the objective function of the given problem.

GAs are closely related to *evolutionary strategies*. Whereas the mutation operator in a GA serves to protect the search from premature loss of information, evolution strategies may incorporate some sort of local search procedure (such as SD) with self adapting parameters involved in the procedure.

---

[4] A new idea worth investigating applies a SA run instead.

For some interesting insights on evolutionary algorithms the reader is referred to [46]. On a very simple scale many algorithms may be coined evolutionary once they are reduced to the following frame:

1. Generate an initial population of individuals
2. While no stopping condition is met do
   a) co-operation
   b) self-adaptation

Self-adaptation refers to the fact that individuals (solutions) evolve independently while co-operation refers to an information exchange among individuals.

Recently it appeared that so-called *scatter search* ideas may establish a link between early ideas from various sides – evolutionary strategies, TS and GAs. Scatter search is designed to operate on a set of points, called reference points, that constitute good solutions obtained from previous solution efforts. The approach systematically generates linear combinations of the reference points to create new points, each of which is mapped into an associated point that yields integer values for discrete variables.

3.5   Ant Systems

One of the recently explored concepts within intelligent search is the *ant system*, a dynamic optimization process reflecting the natural interaction between ants searching for food (see, e.g., [17,100,101]). The ants' ways are influenced by two different kinds of search criteria. The first one is the local visibility of food, i.e., the attractiveness of food in each ant's neighborhood. Additionally, each ant's way through its food space is affected by the other ants' trails as indicators for possibly good directions. The intensity of trails itself is time-dependent: With time going by, parts of the trails 'are gone with the wind', meanwhile the intensity may increase by new and fresh trails. With the quantities of these trails changing dynamically, an autocatalytic optimization process is started forcing the ants' search into most promising regions. This process of interactive learning can easily be modeled for most kinds of optimization problems by using simultaneously and interactively processed search trajectories.

To achieve enhanced performance of the ant system it is useful to hybridize it with a local search component. For a successful application of such an approach to, e.g., the sequential ordering problem see [29].

3.6   Miscellaneous

Target analysis may be viewed as a general learning approach. Given a problem we first explore a set of sample instances and an extensive effort is made to obtain a solution which is optimal or close to optimality. The best solutions obtained will provide some *targets* to be sought within the next part of the approach. For instance, a TS algorithm may resolve the problems with the aim of finding what are the right choices to come to the already known solution (or as close to it as possible). This may give some information on how to choose parameters for other problem instances. A different acronym in this context is *path relinking* which provides a useful means of intensification and diversification. Here new solutions are generated by exploring search trajectories that combine elite solutions, i.e., solutions that have proven to be better than others throughout the search. For references on target analysis and path relinking see, e.g., [37].

Recalling local search based on *data perturbation* the acronym *noising method* may be related to the following approach, too. Given an initial feasible solution, the method performs some data perturbation [99] in order to change the values taken by the objective function of a respective problem to be solved. With this perturbed data some local search iterations may be performed (e.g., following a SD approach). The amount of data perturbation (the noise added) is successively reduced until it reaches zero. The noising method is applied, e.g., in [13] for the clique partitioning problem and in [88] for the Steiner problem in graphs as a hybrid with GRASP.

The key issue in designing *parallel algorithms* is to decompose the execution of the various ingredients of a procedure into processes executable by parallel processors. Opposite to ant systems or GAs meta-heuristics like TS or SA, at first glance, have an intrinsic sequential nature due to the idea of performing the neighborhood search from one solution to the next. However, some effort has been undertaken to define templates for parallel local search (see, e.g., [109,106,15, 104]). Some successful applications are referenced in [2]. The discussion of parallel meta-heuristics has also led to interesting hybrids such as the combination of a population of individual processes, agents, in a cooperative and competitive nature (see, e.g., the discussion of *memetic algorithms* in [71]) with TS.

Of course *neural networks* may be considered as meta-heuristics, although we have not considered them in this paper; see, e.g., [97] for a comprehensive survey on these techniques for combinatorial optimization. Furthermore, we have not considered problems with multiple objectives and corresponding approaches (see, e.g., [91] for some ideas regarding GAs and fuzzy multiobjective optimization).

One of the trends not to be overlooked is that of *soft computing*. It differs from conventional computing in that it is tolerant of imprecision, uncertainty, partial truth, and approximation. The principal constituents of soft computing (sometimes also called computational intelligence) are fuzzy computing but also evolutionary computation [11], neural computing, and probabilistic reasoning, with the latter subsuming parts of machine learning. The concept of *fuzzy sets and systems* has received considerable attention with respect to its general concept as well as applications in various fields especially when models and data are vague or fuzzy (see, e.g., [119] for a survey of concepts and applications). Certainly, meta-heuristics should play an important role in soft computing.

4   Recent Advances on General Frames

A subset of recent advances on meta-heuristics refers to general frames (to explain the behavior and the relationship between various methods) as well as the development of software systems incorporating meta-heuristics (eventually in combination with other methods). Besides other aspects, this takes into consideration that in meta-heuristics it has very often been appropriate to incorporate a certain means of diversification versus intensification to lead the search into new regions of the search space. This requires a meaningful mechanism to detect situations when the search might be trapped in a certain area of the solution space. Therefore, within intelligent search the exploration of memory plays a most important role.

4.1   Adaptive Memory Programming

Adaptive memory programming (AMP) coins a general approach (or even thinking) within heuristic search focusing on exploiting a collection of memory components [35,103]. That is, iteratively constructing (new) solutions based on the exploitation of some sort of memory may be viewed as AMP process, especially when combined with learning mechanisms that helps to adapt the collection and use of the memory. Based on the simple idea of initializing the memory and then iteratively generating new solutions (utilizing the given memory) while updating the memory based on the search, we may subsume various of the above described meta-heuristics as AMP approaches. This also includes the idea of exploiting provisional solutions that are improved by a local search approach.

The performance as well as the efficiency of a heuristic scheme strongly depends on its ability to use AMP techniques providing flexible and variable strategies for types of problems (or special instances of a given problem type) where standard methods fail. Such AMP techniques could be, e.g., dynamic handling of operational restrictions, dynamic move selection formulas, and flexible function evaluations.

Consider, as an example, adaptive memory within TS concepts. Realizing AMP principles depends on which specific TS application is used. For example, the reverse elimination method observes logical interdependencies between moves and infers corresponding tabu restrictions, and therefore makes fuller use of AMP than simple static approaches do.

To discuss the use of AMP in intelligent agent systems, we may also refer to the simple model of ant systems as an illustrative starting point. Ant systems are based on combining local search criteria with information derived from the trails. This follows the AMP requirement for using flexible (dynamic) move selection rules (formulas). However, the basic ant system exhibits some structural inefficiencies when viewed from the perspective of general intelligent agent systems, as no distinction is made between successful and less successful agents, no time-dependent distinction is made, there is no explicit handling of restrictions providing protection against cycling and duplication. Furthermore, there are possible conflicts between the information held in the adaptive memory (*diverging trails*).

4.2   POPMUSIC – Partial Optimization Meta-heuristic under Special Intensification Conditions

A natural way to solve large combinatorial optimization problems is to decompose them into independent sub-problems that are solved with an appropriate procedure. However, such approaches may lead to solutions of moderate quality since the sub-problems might have been created in a somewhat arbitrary fashion. Indeed, it is not easy to find an appropriate way to decompose a problem *a priori*. The basic idea of POPMUSIC is to locally optimize sub-parts of a solution, *a posteriori*, once a solution to the problem is available. These local optimizations are repeated until a local optimum is found. So, POPMUSIC can be seen as a local search working with a special, large neighborhood. While POPMUSIC has been acronymed by [102] other meta-heuristics may be incorporated into the same framework, too (e.g. [96]).

For large optimization problems, it is often possible to see the solutions as composed of parts (or chunks [116], cf. the term vocabulary building). Considering the vehicle routing problem, a part may be a tour (or even a customer). Suppose that a solution can be represented as a set of parts. Moreover, some parts are more in relation with some other parts so that a relatedness measure can be defined between two parts. The central idea of POPMUSIC is to select a so-called *seed part* and a set P of parts that are mostly related with the seed part to form a sub-problem.

Then it is possible to state a local search optimization frame that consists of trying to improve all sub-problems that can be defined, until the solution does not contain a sub-problem that can be improved. In the POPMUSIC frame of [102], P corresponds precisely to seed parts that have been used to define sub-problems that have been unsuccessfully optimized. Once P contains all the parts of the complete solution, then all sub-problems have been examined without success and the process stops.

Basically, the technique is a gradient method that starts from a given initial solution and stops in a local optimum relative to a large neighborhood structure. To summarize, both, POPMUSIC as well as AMP may serve as a general frame encompassing various other approaches.

4.3   Constraint Programming

Constraint programming (CP) is a paradigm for representing and solving a wide variety of problems expressed by means of variables, their domains, and constraints on the variables (see, e.g., [105,50,45,55] for some surveys from slightly different perspectives). Usually CP models are solved using depth-first search and branch and bound. Naturally, these concepts can be complemented by local search concepts and meta-heuristics. This idea is followed by several authors; see, e.g., [16] for TS and guided local search hybrids. Interestingly, commonalities with our POPMUSIC approach exist, as can be deduced from [89,80].

Of course, the treatment of this topic is by no means complete and various ideas have been developed (see, e.g., those from [74,75] regarding local search within CP). Especially, the use of higher level or global constraints, i.e., those which use domain specific knowledge based on suitable problem representations, may influence the search efficiency. Another idea is to transform a greedy heuristic into a search algorithm by branching only in a few (i.e., limited number) cases when the choice criterion of the heuristic observes some borderline case or where the choice is least compelling, respectively. This approach may be called *limited discrepancy search* (see, e.g., [44,12]).

Independently from the CP concept one may investigate hybrids of branch and bound and meta-heuristics, e.g., for deciding upon branching variables or search paths to be followed within a branch and bound tree (see, e.g., [117] for an application of reactive TS). Here we may also use the term *cooperative solver*. Combining TS with Lagrangean relaxation and Lagrangean decomposition is proposed in [40].

4.4   Propositional Satisfiability

While the adaptive memory programming and the POPMUSIC frames are more operations research related, a considerable amount of work especially within the computer science community refers to local search for propositional satisfiability (SAT). For given sets of variables and clauses SAT asks for the existence of an assignment to the variables that satisfies all clauses. While SAT is a decision problem the problems MAX-SAT and weighted MAX-SAT are related problems, which ask for a propositional variable assignment that maximizes the number of satisfied and the weighted sum of satisfied (weighted) clauses, respectively.

The classic local search strategies for SAT are GSAT [95] and modifications. Moves are simple changes of variable assignments (flip moves) chosen in a greedy randomized fashion by means of maximizing the number of satisfied clauses. One of the modifications is WalkSAT [94] where some noise (data perturbation regarding randomly chosen initial values assigned to the variables) is added to the strategy. Throughout the years various modifications for local search for SAT have been investigated (see, e.g., [60,61,41,112,92]). A very comprehensive empirical evaluation of GSAT and WalkSAT can be found in [52]. To increase the number of feasible clauses often so-called repair algorithms are exploited. For a thought provoking connection of such an algorithm to Lagrangian multiplier methods see [14].

The basic idea regarding the solution of a great variety of problems is to describe them by means of instances of SAT or MAX-SAT (SAT encoded problems) and to apply solution methods available for these problems (see, e.g., [56]). While many problems formulated under this paradigm have been solved successfully following such approaches, however, this is not necessarily the case for some classical combinatorial optimization problems such as the Steiner problem in graphs.

4.5   Optimization Software Libraries

Systems may be developed to support the implementation of local search and meta-heuristics. Such algorithms can be supported by (high-level) modeling languages that reduce their development time substantially while preserving most

of the efficiency of special purpose implementations. For instance, the design of
LOCALIZER [69,70] may be viewed as such an approach with several extensions
being contemplated, including additional support for some meta-heuristics and
the integration of consistency techniques.

While there are some well-known approaches for reusable software in the field
of exact optimization,[5] there is, as far as we know, almost no ready-to-use and
well-documented component library in the field of local search based heuristics
and meta-heuristics. Besides the discussion below, we would like to point out
the class library for heuristic optimization of [115], which includes both local
search based methods and genetic algorithms, the frameworks for local search
heuristics of [5,58], and the class library described in [26].

Here we refer to another successful approach, i.e. HOTFRAME of [24,53],
a Heuristic OpTimization FRAMEwork implemented in C++, which provides
adaptable components incorporating different meta-heuristics and an architec-
tural description of the collaboration among these components and problem-
specific complements. Typical application-specific concepts are treated as objects or classes: problems, solutions, neighbors, solution attributes and move
attributes. On the other side, meta-heuristic concepts such as different methods
described above and their building-blocks such as tabu criteria or diversification strategies are also treated as objects. HOTFRAME uses genericity as the
primary mechanism to make these objects adaptable. That is, common behavior
of meta-heuristics is factored out and grouped in generic classes, applying static
type variation. Meta-heuristics template classes are parameterized by aspects
such as solution spaces and neighborhood structures.

HOTFRAME defines an architecture for the interplay between heuristic classes
and application specific classes, and provides several such classes, which implement classic methods that are applicable to almost arbitrary problem types,
solution spaces and neighborhood structures:

- A simple local search frame, which can be instantiated by, e.g., different
  neighbor selection rules resulting in methods such as SD or iterated local
  search.
- Several SA and threshold accepting methods, which can be customized by
  different acceptance criteria, cooling schedules, and reheating schemes.
- A classic TS frame, which is based on treating both the tabu criterion and
  the diversification strategy as parameters. Move attributes, trajectory information and tabu criteria (taking into account information about the search
  history such as memory, performed moves, and traversed solutions) can be
  flexibly defined. Static, strict and reactive tabu criteria are built-in. Other
  TS methods can be implemented exploiting existing basic building-blocks.
- Candidate list templates can be used as filters in connection with various
  neighborhood structures and heuristics. On the other hand, VNS may be
  used to enlarge neighborhoods.

---

[5] Consider, e.g., mathematical programming software like CPLEX or class-libraries
for constraint programming such as ILOG Solver (http://www.ilog.com/).

- A general evolutionary algorithm template, which is customizable with respect to arbitrary solution representations, selection and replacement strategies, recombination (not necessarily restricted to two parents, cf. the idea of scatter search) and mutation operators.
- Different means for determining initial feasible solutions including the pilot method are incorporated.

All heuristics are implemented in a consistent way, which facilitates an easy embedding of arbitrary methods into application systems or as parts of more advanced/hybrid methods. Both, new meta-heuristics and new applications, can be added into the framework. HOTFRAME includes built-in support for solution spaces representable by binary vectors or permutations, in connection with corresponding standard neighborhood structures, solution and move attributes, and recombination operators. Otherwise, the user may derive specialized classes from suitable built-in classes or implement corresponding classes from scratch according to defined interfaces.

For example, to apply HOTFRAME for some problem with a permutation like solution space, one can derive a specific class from a general class that represents solutions defined by a permutation, and implement the computation of objective function values. Then, all predefined meta-heuristics can be applied by reusing predefined classes that represent, e.g., neighborhood structures (such as shift or swap) or move attributes. Of course, to efficiently apply methods based on the local search paradigm, the evaluation of moves should be computed in an adaptive way.

5   Conclusions

Over the last decade meta-heuristics have become a substantial part of the heuristics stockroom with various applications in science and, even more important, in practice. Meta-heuristics have become part of textbooks, e.g. in operations research, and a wealth of monographs (see, e.g., [108,37,68,90]) is available. Specialized conferences are devoted to the topic (see, e.g., [77,111]).[6] Even ready to use systems such as class libraries and frameworks are under development, although usually restricted to be used by the *knowledgeable* user. In this respect the development of an adoption path taking into account the needs of different users at different levels is still a challenge, see [25] for some ideas. Whether concepts like case based reasoning may help in this respect is still open [39].

Most important are general frames. Adaptive memory programming, an intelligent interplay of intensification and diversification (such as ideas from POP-MUSIC), and the connection to powerful exact algorithms as subroutines for handable subproblems are avenues to be followed.

---

[6] Interested readers may even contact relevant interest groups. See, e.g., modern-heuristics@mailbase.ac.uk and evolutionary-computing@mailbase.ac.uk as well as http://tew.ruca.ua.ac.be/eume.

From a theoretical point of view, the use of most meta-heuristics has not yet been fully justified. While convergence results regarding solution quality exist for most meta-heuristics once appropriate probabilistic assumptions are made, these turn out not to be very helpful in practice as usually a disproportionate computation time is required to achieve these results (usually convergence is achieved for the computation time tending to infinity, with a few exceptions, e.g., for the reverse elimination method within tabu search or the pilot method where optimality can be achieved with a finite, but exponential number of steps in the worst case). Furthermore, we have to admit that theoretically one may argue that none of the described meta-heuristics is *on average* better than any other; there is *no free lunch* [114].

From an empirical standpoint it would be most interesting to know which algorithms perform best under various criteria for different classes of problems. Unfortunately, this theme is out of reach as long as we do not have any well accepted standards regarding the testing and comparison of different methods.

While most papers on meta-heuristics claim to provide 'high quality' results based on some sort of measure, we still believe that there is a great deal of room for improvement in testing existing as well as new approaches from an empirical point of view (see, e.g., [7,49,66]). In a dynamic research process numerical results provide the basis for systematically developing efficient algorithms. The essential conclusions of finished research and development processes should always be substantiated (i.e., empirically and, if necessary, statistically proven) by numerical results based on an appropriate empirical test cycle.

Usually the ways of preparing, performing and presenting experiments and their results are significantly different. The failing of a generally accepted standard for testing and reporting on the testing, or at least a corresponding guideline for designing experiments, unfortunately leads to the following observation: Parts of results can be used only in a restricted way, e.g., because relevant data are missing, wrong environmental settings are used, or simply results are glossed over. In the worst case non-sufficiently prepared experiments provide results that are unfit for further use, i.e., any generalized conclusion is out of reach. Future algorithm research needs to provide effective methods for analyzing the performance of, e.g., heuristics in a more scientifically founded way (see, e.g., [98,113, 51] for some steps into this direction).

A final aspect that deserves special consideration is to investigate the use of information within different meta-heuristics. While the adaptive memory programming frame provides a very good entry into this area, this still provides an interesting opportunity to link artificial intelligence with operations research concepts.

References

1. E.H.L. Aarts and J.K. Lenstra, editors. *Local Search in Combinatorial Optimization*. Wiley, Chichester, 1997.
2. E.H.L. Aarts and M. Verhoeven. Local search. In M. Dell'Amico, F. Maffioli, and S. Martello, editors, *Annotated Bibliographies in Combinatorial Optimization*, pages 163–180. Wiley, Chichester, 1997.

3. R.K. Ahuja, O. Ergun, J.B. Orlin, and A.B. Punnen. A survey of very large-scale neighborhood search techniques. Working paper, Sloan School of Management, MIT, 1999.

4. E.J. Anderson, C.A. Glass, and C.N. Potts. Machine scheduling. In E.H.L. Aarts and J.K. Lenstra, editors, *Local Search in Combinatorial Optimization*, pages 361–414. Wiley, Chichester, 1997.

5. A.A. Andreatta, S.E.R. Carvalho, and C.C. Ribeiro. An object-oriented framework for local search heuristics. In *Proceedings of the 26th Conference on Technology of Object-Oriented Languages and Systems (TOOLS USA'98)*, pages 33–45. IEEE, Piscataway, 1998.

6. T. Bäck, D.B. Fogel, and Z. Michalewicz, editors. *Handbook of Evolutionary Computation*. Institute of Physics Publishing, Bristol, 1997.

7. R.S. Barr, B.L. Golden, J.P. Kelly, M.G.C. Resende, and W.R. Stewart. Designing and reporting on computational experiments with heuristic methods. *Journal of Heuristics*, 1:9–32, 1995.

8. R. Battiti. Reactive search: Toward self-tuning heuristics. In V.J. Rayward-Smith, I.H. Osman, C.R. Reeves, and G.D. Smith, editors, *Modern Heuristic Search Methods*, pages 61–83. Wiley, Chichester, 1996.

9. D.P. Bertsekas and D.A. Castanon. Rollout algorithms for stochastic scheduling problems. *Journal of Heuristics*, 5:89–108, 1999.

10. D.P. Bertsekas, J.N. Tsitsiklis, and C. Wu. Rollout algorithms for combinatorial optimization. *Journal of Heuristics*, 3:245–262, 1997.

11. J.C. Bezdek. What is Computational Intelligence. In J.M. Zurada, R.J. Marks II, and C.J. Robinson, editors, *Computational Intelligence: Imitating Life*, pages 1–12. IEEE Press, New York, 1994.

12. Y. Caseau, F. Laburthe, and G. Silverstein. A meta-heuristic factory for vehicle routing problems. In J. Jaffar, editor, *Principles and Practice of Constraint Programming – CP '99*, Lecture Notes in Computer Science 1713, pages 144–158. Springer, Berlin, 1999.

13. I. Charon and O. Hudry. The noising method: A new method for combinatorial optimization. *Operations Research Letters*, 14:133–137, 1993.

14. K.M.F. Choi, J.H.M. Lee, and P.J. Stuckey. A Lagrangian reconstruction of GENET. *Artificial Intelligence*, 123:1–39, 2000.

15. T.G. Crainic, M. Toulouse, and M. Gendreau. Toward a taxonomy of parallel tabu search heuristics. *INFORMS Journal on Computing*, 9:61–72, 1997.

16. B. de Backer, V. Furnon, P. Shaw, P. Kilby, and P. Prosser. Solving vehicle routing problems using constraint programming and metaheuristics. *Journal of Heuristics*, 6:501–523, 2000.

17. M. Dorigo, V. Maniezzo, and A. Colorni. Ant system: Optimization by a colony of cooperating agents. *IEEE Transactions on Systems, Man and Cybernetics*, B - 26:29–41, 1996.

18. K.A. Dowsland. Simulated annealing. In C. Reeves, editor, *Modern Heuristic Techniques for Combinatorial Problems*, pages 20–69. Halsted, Blackwell, 1993.

19. G. Dueck and T. Scheuer. Threshold accepting: a general purpose optimization algorithm appearing superior to simulated annealing. *Journal of Computational Physics*, 90:161–175, 1990.

20. C.W. Duin and S. Voß. Steiner tree heuristics - a survey. In H. Dyckhoff, U. Derigs, M. Salomon, and H.C. Tijms, editors, *Operations Research Proceedings 1993*, pages 485–496, Berlin, 1994. Springer.

21. C.W. Duin and S. Voß. The pilot method: A strategy for heuristic repetition with application to the Steiner problem in graphs. *Networks*, 34:181–191, 1999.

22. T.A. Feo, K. Venkatraman, and J.F. Bard. A GRASP for a difficult single machine scheduling problem. *Computers & Operations Research*, 18:635–643, 1991.
23. A. Fink and S. Voß. Applications of modern heuristic search methods to continuous flow-shop scheduling problems. Working paper, Technische Universität Braunschweig, Germany, 1999.
24. A. Fink and S. Voß. Generic metaheuristics application to industrial engineering problems. *Computers & Industrial Engineering*, 37:281–284, 1999.
25. A. Fink, S. Voß, and D.L. Woodruff. An adoption path for intelligent heuristic search componentware. In E. Rolland and N.S. Umanath, editors, *Proceedings of the 4th INFORMS Conference on Information Systems and Technology*, pages 153–168. INFORMS, Linthicum, 1999.
26. C. Fleurent and J.A. Ferland. Object-oriented implementation of heuristic search methods for graph coloring, maximum clique, and satisfiability. In D.S. Johnson and M.A. Trick, editors, *Cliques, Coloring, and Satisfiability: Second DIMACS Implementation Challenge*, volume 26 of *DIMACS Series in Discrete Mathematics and Theoretical Computer Science*, pages 619–652. AMS, Princeton, 1996.
27. D.B. Fogel. On the philosophical differences between evolutionary algorithms and genetic algorithms. In D.B. Fogel and W. Atmar, editors, *Proceedings of the Second Annual Conference on Evolutionary Programming*, pages 23–29. Evolutionary Programming Society, La Jolla, 1993.
28. D.B. Fogel. *Evolutionary Computation: Toward a New Philosophy of Machine Intelligence*. IEEE Press, New York, 1995.
29. L.M. Gambardella and M. Dorigo. An ant colony system hybridized with a new local search for the sequential ordering problem. *INFORMS Journal on Computing*, 12:237–255, 2000.
30. M.R. Garey and D.S. Johnson. *Computers and Intractability, A Guide to the Theory of $\mathcal{NP}$-Completeness*. Freeman, New York, 1979.
31. F. Glover. Heuristics for integer programming using surrogate constraints. *Decision Sciences*, 8:156–166, 1977.
32. F. Glover. Future paths for integer programming and links to artificial intelligence. *Computers & Operations Research*, 13:533–549, 1986.
33. F. Glover. Tabu search – Part II. *ORSA Journal on Computing*, 2:4–32, 1990.
34. F. Glover. Scatter search and star-paths: beyond the genetic metaphor. *OR Spektrum*, 17:125–137, 1995.
35. F. Glover. Tabu search and adaptive memory programming – Advances, applications and challenges. In R.S. Barr, R.V. Helgason, and J.L. Kennington, editors, *Interfaces in Computer Science and Operations Research: Advances in Metaheuristics, Optimization, and Stochastic Modeling Technologies*, pages 1–75. Kluwer, Boston, 1997.
36. F. Glover, editor. *Tabu Search Methods for Optimization*. European Journal of Operational Research 106:221–692. Elsevier, Amsterdam, 1998.
37. F. Glover and M. Laguna. *Tabu Search*. Kluwer, Boston, 1997.
38. D.E. Goldberg. *Genetic Algorithms in Search, Optimization, and Machine Learning*. Addison-Wesley, Reading, 1989.
39. S. Grolimund and J.-G. Ganascia. Driving tabu search with case-based reasoning. *European Journal of Operational Research*, 103:326–338, 1997.
40. T. Grünert. Lagrangean tabu search. In C.C. Ribeiro, editor, *Third Metaheuristics International Conference: Extended Abstracts*, pages 263–267, 1999.
41. J. Gu. The Multi-SAT algorithm. *Discrete Applied Mathematics*, 96–97:111–126, 1999.

42. P. Hansen and N. Mladenović. An introduction to variable neighborhood search. In S. Voß, S. Martello, I.H. Osman, and C. Roucairol, editors, *Meta-Heuristics: Advances and Trends in Local Search Paradigms for Optimization*, pages 433–458. Kluwer, Boston, 1999.

43. J.P. Hart and A.W. Shogan. Semi-greedy heuristics: An empirical study. *Operations Research Letters*, 6:107–114, 1987.

44. W. Harvey and M. Ginsberg. Limited discrepancy search. In *Proceedings of the 14th IJCAI*, pages 607–615, San Mateo, 1995. Morgan Kaufmann.

45. S. Heipcke. Comparing constraint programming and mathematical programming approaches to discrete optimisation – the change problem. *Journal of the Operational Research Society*, 50:581–595, 1999.

46. A. Hertz and D. Kobler. A framework for the description of evolutionary algorithms. *European Journal of Operational Research*, 126:1–12, 2000.

47. F. Hoffmeister and T. Bäck. Genetic algorithms and evolution strategies: Similarities and differences. In H.-P. Schwefel and R. Männer, editors, *Parallel Problem Solving from Nature – PPSN I*, Lecture Notes in Computer Science 496, pages 455–469. Springer, Berlin, 1991.

48. J.H. Holland. *Adaptation in Natural and Artificial Systems*. The University of Michigan Press, Ann Arbor, 1975.

49. J.N. Hooker. Testing heuristics: We have it all wrong. *Journal of Heuristics*, 1:33–42, 1995.

50. J.N. Hooker. Constraint satisfaction methods for generating valid cuts. In D.L. Woodruff, editor, *Advances in Computational and Stochastic Optimization, Logic Programming, and Heuristic Search*, pages 1–30. Kluwer, Boston, 1998.

51. H.H. Hoos and T. Stützle. Evaluating Las Vegas algorithms - Pitfalls and remedies. In *Proceedings of UAI-98*, pages 238–245. 1998.

52. H.H. Hoos and T. Stützle. Local search algorithms for SAT. *Journal of Automated Reasoning*, 24:421–481, 2000.

53. HOTFRAME/Heuristic Optimization Framework. http://www.winforms.phil.tu-bs.de/research/hotframe.htm, 2000.

54. L. Ingber. Adaptive simulated annealing (ASA): Lessons learned. *Control and Cybernetics*, 25:33–54, 1996.

55. J. Jaffar, editor. *Principles and Practice of Constraint Programming – CP '99*. Lecture Notes in Computer Science 1713. Springer, Berlin, 1999.

56. Y. Jiang, H. Kautz, and B. Selman. Solving problems with hard and soft constraints using a stochastic algorithm for MAX-SAT. Technical report, AT&T Bell Laboratories, 1995.

57. D.S. Johnson, C.R. Aragon, L.A. McGeoch, and C. Schevon. Optimization by simulated annealing: An experimental evaluation; part i, graph partitioning. *Operations Research*, 37:865–892, 1989.

58. M.S. Jones, G.P. McKeown, and V.J. Rayward-Smith. Distribution, cooperation, and hybridization for combinatorial optimization. Technical report, University of East Anglia, Norwich, 2000.

59. A.B. Kahng and G. Robins. A new class of iterative Steiner tree heuristics with good performance. *IEEE Transactions on Computer-Aided Design*, 11:893–902, 1992.

60. H. Kautz and B. Selman. Pushing the envelope: Planning, propositional logic, and stochastic search. In *Proceedings of the 13th National Conference on Artificial Intelligence (AAAI-96)*, pages 1194–1201. 1996.

61. H. Kautz, B. Selman, and Y. Jiang. General stochastic approach to solving problems with hard and soft constraints. In D. Gu, J. Du, and P. Pardalos, editors, *The Satisfiability Problem: Theory and Applications*, DIMACS Series in Discrete Mathematics and Theoretical Computer Science 35, pages 573–586. AMS, Providence, 1997.

62. S. Kirkpatrick, C.D. Gelatt Jr., and M.P. Vecchi. Optimization by simulated annealing. *Science*, 220:671–680, 1983.

63. G. Laporte and I.H. Osman, editors. *Metaheuristics in Combinatorial Optimization*. Annals of Operations Research 63. Baltzer, Amsterdam, 1996.

64. J.L. Lauriere. A language and a program for stating and solving combinatorial problems. *Artificial Intelligence*, 10:29–127, 1978.

65. S. Lin and B.W. Kernighan. An effective heuristic algorithm for the traveling-salesman problem. *Operations Research*, 21:498–516, 1973.

66. C. McGeoch. Toward an experimental method for algorithm simulation. *INFORMS Journal on Computing*, 8:1–15, 1996.

67. Z. Michalewicz. *Genetic Algorithms + Data Structures = Evolution Programs*. Springer, Berlin, 3rd edition, 1999.

68. Z. Michalewicz and D.B. Fogel. *How to Solve It: Modern Heuristics*. Springer, Berlin, 1999.

69. L. Michel and P. van Hentenryck. LOCALIZER: A modeling language for local search. *INFORMS Journal on Computing*, 11:1–14, 1999.

70. L. Michel and P. van Hentenryck. Localizer. *Constraints*, 5:43–84, 2000.

71. P. Moscato. An introduction to population approaches for optimization and hierarchical objective functions: A discussion on the role of tabu search. *Annals of Operations Research*, 41:85–121, 1993.

72. H. Mühlenbein. Genetic algorithms. In E.H.L. Aarts and J.K. Lenstra, editors, *Local Search in Combinatorial Optimization*, pages 137–171. Wiley, Chichester, 1997.

73. H. Müller-Merbach. Heuristics and their design: a survey. *European Journal of Operational Research*, 8:1–23, 1981.

74. A. Nareyek. Using Global Constraints for Local Search. In E. C. Freuder and R. J. Wallace, editors, *Constraint Programming and Large Scale Discrete Optimization*, DIMACS Volume 57, pages 9–28. American Mathematical Society Publications, Providence, 2001.

75. A. Nareyek. Beyond the Plan-Length Criterion. In A. Nareyek, editor, *Local Search for Planning and Scheduling*. Springer LNAI 2048, Berlin, 2001. (this volume)

76. I.H. Osman. Heuristics for the generalized assignment problem: simulated annealing and tabu search approaches. *OR Spektrum*, 17:211–225, 1995.

77. I.H. Osman and J.P. Kelly, editors. *Meta-Heuristics: Theory and Applications*. Kluwer, Boston, 1996.

78. I.H. Osman and G. Laporte. Metaheuristics: A bibliography. *Annals of Operations Research*, 63:513–623, 1996.

79. J. Pearl. *Heuristics: Intelligent Search Strategies for Computer Problem Solving*. Addison-Wesley, Reading, 1984.

80. G. Pesant and M. Gendreau. A constraint programming framework for local search methods. *Journal of Heuristics*, 5:255–279, 1999.

81. E. Pesch and F. Glover. TSP ejection chains. *Discrete Applied Mathematics*, 76:165–182, 1997.

82. E. Pesch and S. Voß, editors. *Applied Local Search*. OR Spektrum 17:55–225. Springer, Berlin, 1995.

83. G. Polya. *How to solve it*. Princeton University Press, Princeton, 1945.
84. C. Potts and S. van de Velde. Dynasearch - iterative local improvement by dynamic programming. Technical report, University of Twente, 1995.
85. V.J. Rayward-Smith, editor. *Applications of Modern Heuristic Methods*. Waller, Henley-on-Thames, 1995.
86. V.J. Rayward-Smith, I.H. Osman, C.R. Reeves, and G.D. Smith, editors. *Modern Heuristic Search Methods*. Wiley, Chichester, 1996.
87. C.R. Reeves, editor. *Modern Heuristic Techniques for Combinatorial Problems*. Blackwell, Oxford, 1993.
88. C.C. Ribeiro, E. Uchoa, and R.F. Werneck. A hybrid GRASP with perturbations for the Steiner problem in graphs. Technical report, Department of Computer Science, Catholic University of Rio de Janeiro, 2000.
89. L.-M. Rousseau, M. Gendreau, and G. Pesant. Using constraint-based operators to solve the vehicle routing problem with time windows. Technical report, CRT, University of Montreal, Canada, 2000.
90. S.M. Sait and H. Youssef. *Iterative Computer Algorithms with Applications in Engineering: Solving Combinatorial Optimization Problems*. IEEE Computer Society Press, Los Alamitos, 1999.
91. M. Sakawa and T. Shibano. Multiobjective fuzzy satisficing methods for 0-1 knapsack problems through genetic algorithms. In W. Pedrycz, editor, *Fuzzy Evolutionary Computation*, pages 155–177. Kluwer, Boston, 1997.
92. D. Schuurmans and F. Southey. Local search characteristics of incomplete SAT procedures. In *Proceedings of the 17th National Conference on Artificial Intelligence (AAAI-2000)*, pages 297–302. 2000.
93. H.-P. Schwefel and T. Bäck. Artificial evolution: How and why? In D. Quagliarella, J. Périaux, C. Poloni, and G. Winter, editors, *Genetic Algorithms and Evolution Strategy in Engineering and Computer Science: Recent Advances and Industrial Applications*, pages 1–19. Wiley, Chichester, 1998.
94. B. Selman, H. Kautz, and B. Cohen. Noise strategies for improving local search. In *Proceedings of the 11th National Conference on Artificial Intelligence (AAAI-94)*, pages 337–343. 1994.
95. B. Selman, H. Levesque, and D. Mitchell. A new method for solving hard satisfiability problems. In *Proceedings of the 9th National Conference on Artificial Intelligence (AAAI-92)*, pages 440–446. 1992.
96. P. Shaw. Using constraint programming and local search methods to solve vehicle routing problems. Working paper, ILOG S.A., Gentilly, France, 1998.
97. K. Smith. Neural networks for combinatorial optimisation: A review of more than a decade of research. *INFORMS Journal on Computing*, 11:15–34, 1999.
98. L. Sondergeld. *Performance Analysis Methods for Heuristic Search Optimization with an Application to Cooperative Agent Algorithms*. Shaker, Aachen, 2001.
99. R.H. Storer, S.D. Wu, and R. Vaccari. Problem and heuristic space search strategies for job shop scheduling. *ORSA Journal on Computing*, 7:453–467, 1995.
100. T. Stützle and H. Hoos. The max-min ant system and local search for combinatorial optimization problems. In S. Voss, S. Martello, I.H. Osman, and C. Roucairol, editors, *Meta-Heuristics: Advances and Trends in Local Search Paradigms for Optimization*, pages 313–329. Kluwer, Boston, 1999.
101. E. Taillard. An introduction to ant systems. In M. Laguna and J.L. Gonzalez-Velarde, editors, *Computing Tools for Modeling, Optimization and Simulation*, pages 131–144. Kluwer, Boston, 2000.
102. E. Taillard and S. Voß. Popmusic. Working paper, University of Applied Sciences of Western Switzerland, 1999.

103. E.D. Taillard, L.M. Gambardella, M. Gendreau, and J.-Y. Potvin. Adaptive memory programming: A unified view of meta-heuristics. Technical Report IDSIA-19-98, Istituto Dalle Molle di Studi sull'Intelligenza Artificiale, Lugano, 1998.
104. R.J.M. Vaessens, E.H.L. Aarts, and J.K. Lenstra. A local search template. *Computers & Operations Research*, 25:969–979, 1998.
105. P. van Hentenryck. Constraint solving for combinatorial search problems: A tutorial. In U. Montanari and F. Rossi, editors, *Principles and Practice of Constraint Programming – CP '95*, Lecture Notes in Computer Science 976, pages 564–587. Springer, Berlin, 1995.
106. M.G.A. Verhoeven and E.H.L. Aarts. Parallel local search techniques. *Journal of Heuristics*, 1:43–65, 1995.
107. R.V.V. Vidal, editor. *Applied Simulated Annealing*. Lecture Notes in Economics and Mathematical Systems 396. Springer, Berlin, 1993.
108. S. Voß. *Intelligent Search*. Manuscript, TU Darmstadt, 1993.
109. S. Voß. Tabu search: applications and prospects. In D.-Z. Du and P. Pardalos, editors, *Network Optimization Problems*, pages 333–353. World Scientific, Singapore, 1993.
110. S. Voß. Observing logical interdependencies in tabu search: Methods and results. In V.J. Rayward-Smith, I.H. Osman, C.R. Reeves, and G.D. Smith, editors, *Modern Heuristic Search Methods*, pages 41–59, Chichester, 1996. Wiley.
111. S. Voß, S. Martello, I.H Osman, and C. Roucairol, editors. *Meta-Heuristics: Advances and Trends in Local Search Paradigms for Optimization*. Kluwer, Boston, 1999.
112. J.P. Walser. *Integer Optimization by Local Search*. Lecture Notes in Artificial Intelligence 1637. Springer, Berlin, 1999.
113. D. Whitley, S. Rana, J. Dzubera, and K.E. Mathias. Evaluating evolutionary algorithms. *Artificial Intelligence*, 85:245–276, 1996.
114. D.H. Wolpert and W.G. Macready. No free lunch theorems for optimization. *IEEE Transactions on Evolutionary Computation*, 1:67–82, 1997.
115. D.L. Woodruff. A class library for heuristic search optimization. *INFORMS Computer Science Technical Section Newsletter*, 18 (2):1–5, 1997.
116. D.L. Woodruff. Proposals for chunking and tabu search. *European Journal of Operational Research*, 106:585–598, 1998.
117. D.L. Woodruff. A chunking based selection strategy for integrating meta-heuristics with branch and bound. In S. Voß, S. Martello, I.H. Osman, and C. Roucairol, editors, *Meta-Heuristics: Advances and Trends in Local Search Paradigms for Optimization*, pages 499–511. Kluwer, Boston, 1999.
118. S.H. Zanakis, J.R. Evans, and A.A. Vazacopoulos. Heuristic methods and applications: a categorized survey. *European Journal of Operational Research*, 43:88–110, 1989.
119. H.-J. Zimmermann. *Fuzzy Set Theory and its Applications*. Kluwer, Boston, 2nd edition, 1991.

Solving the Sports League Scheduling Problem with Tabu Search

Jean-Philippe Hamiez[1] and Jin-Kao Hao[2]

[1] LGI2P / Ecole des Mines d'Alès – EERIE
Parc Scientifique Georges Besse – 30035 Nîmes Cedex 01 (France)
hamiez@site-eerie.ema.fr
[2] LERIA / Université d'Angers – 2, Bd. Lavoisier – 49045 Angers Cedex 01 (France)
Jin-Kao.Hao@univ-angers.fr

Abstract. In this paper we present a tabu approach for a version of the Sports League Scheduling Problem. The approach adopted is based on a formulation of the problem as a Constraint Satisfaction Problem (CSP). Tests were carried out on problem instances of up to 40 teams representing 780 integer variables with 780 values per variable. Experimental results show that this approach outperforms some existing methods and is one of the most promising methods for solving problems of this type.

1 Introduction

Many sports leagues (e.g., soccer, hockey, basketball) must deal with scheduling problems for tournaments. These scheduling problems contain in general many conflicting constraints to satisfy and different objectives to optimize, like minimization of traveling distance [2], only one match per team and per day, stadium unavailability at particular dates, minimum number of days between a home match and its corresponding away match, etc. Generating satisfactory schedules with respect to these conditions and objectives is therefore a very difficult problem to solve.

Many studies have been carried out to try to solve these problems with a variety of approaches and varying degrees of success: integer linear programming [7][12], constraint programming [10][17], local search (simulated annealing [19], tabu search [23], hybrid approach [4]).

This paper deals with a specific Sports League Scheduling Problem (SLSP) described by K. McAloon, C. Tretkoff and G. Wetzel in [11]. After having obtained poor results in integer linear programming tests, they experimented with constraint programming, an approach that produced better results. Finally, with a basic local search algorithm, they produced the same results as ILOG Solver does, but with less computing time.

C.P. Gomes, B. Selman and H.A. Kautz [9] obtained better results than those of McAloon et al. using constraint programming. With a randomized version of a deterministic complete search they solved problems involving a greater number of teams.

J.C. Régin proposed two approaches with constraint programming for the SLSP [15][16]. The first one, using powerful filtering algorithms [3][13][14], produced

A. Nareyek (Ed.): Local Search for Planning and Scheduling, LNAI 2148, pp. 24–36, 2001.
© Springer-Verlag Berlin Heidelberg 2001

better results than those of McAloon et al. and those of Gomes et al. in terms of execution time and robustness, since it solved problem instances of a larger size. Using a second approach, he transformed the SLSP into an equivalent problem by adding an implicit constraint. With a new heuristic and specific filtering algorithms, he improved on his own results.

Finally, let us mention the work done by G. Wetzel and F. Zabatta [22]: using multiple threads on a 14 processor Sun system they obtained better results than the first approach of Régin. They were, however, not able to solve problems as large as those which Régin solved with his second approach.

The goal of this study is to propose an advanced local search approach based on tabu search (TS) [8] for the SLSP. References of the study are results presented in [11], [15] and [16].

The paper begins by formally describing the SLSP (section 2). After modeling the problem as a constraint satisfaction problem (CSP) [20] (section 3), we present our tabu algorithm (section 4) and compare its results with those of [11], [15] and [16] (section 5). Before concluding, we discuss some observations made during this work (section 6).

2   Problem Description

In the rest of the paper, we will deal with the following constraints and definitions, the same as in [11]:

- There are |T| teams (|T| even), where T is the set of all teams. All teams play each other exactly once (half competition);
- The season lasts |T| - 1 weeks;
- Every team plays one game in every week of the season;
- There are |T| / 2 periods and, each week, one game is scheduled in every period;
- No team plays more than twice in the same period over the course of the season.

The problem then is to schedule the tournament with respect to all these constraints.

Table 1 below shows an example of a valid schedule for |T| = 8 teams labeled from 1 to 8; there are 7 weeks and 4 periods.

Table 1. Example of a valid schedule for 8 teams

		Weeks						
		1	2	3	4	5	6	7
Periods	1	1 vs 2	1 vs 3	5 vs 8	4 vs 7	4 vs 8	2 vs 6	3 vs 5
	2	3 vs 4	2 vs 8	1 vs 4	6 vs 8	2 vs 5	1 vs 7	6 vs 7
	3	5 vs 6	4 vs 6	2 vs 7	1 vs 5	3 vs 7	3 vs 8	1 vs 8
	4	7 vs 8	5 vs 7	3 vs 6	2 vs 3	1 vs 6	4 vs 5	2 vs 4

As shown in Table 1, a configuration may be represented as a two-dimensional array with weeks in columns and periods in rows. Each column satisfies the cardinality constraint: each team appears exactly once, i.e., all the teams are different. In each row, no team appears more than twice. There is also a global constraint on the array: each match only appears once, i.e., all matches are different.

3 Problem Formulation

To represent the SLSP we follow the constraint programming approach: we consider it as a constraint satisfaction problem.

3.1 Constraint Satisfaction Problem - CSP

A constraint satisfaction problem [20] is defined by a triplet (X, D, C) with:
- A finite set X of n variables: $X = \{X_1, \ldots, X_n\}$;
- A set D of associated domains: $D = \{D_1, \ldots, D_n\}$. Each domain D_i specifies the finite set of possible values of the variable X_i;
- A finite set C of p constraints: $C = \{C_1, \ldots, C_p\}$. Each constraint is defined for a set of variables and specifies which combinations of values are compatible for these variables.

Given such a triplet, the problem is to generate a complete assignment of the values to the variables, which satisfies all the constraints: such an assignment is said to be consistent. Since the set of all assignments (not necessarily consistent) is defined by the Cartesian product $D_1 \times \ldots \times D_n$ of the domains, solving a CSP means to determine a particular assignment among a potentially huge number of possible assignments.

The CSP, as it has been formalized, is a powerful and general model. In fact, it can be used to conveniently model some well-known problems such as k-coloring and satisfiability, as well as many practical applications relating to resource assignment, planning or timetabling.

3.2 Formulation of the SLSP as a CSP

We will use the following notations to represent the SLSP as a constraint satisfaction problem:
- P: set of periods, $|P| = |T| / 2$;
- W: set of weeks, $|W| = |T| - 1$;
- t_n: team number n, $t_n \in T$, $1 \leq n \leq |T|$;
- $x(t_m, t_n)$: schedule of the match t_m vs. t_n. Values of this variable type are of $(p_{m,n}, w_{m,n})$ pattern, meaning that the match is scheduled in period $p_{m,n}$ and week $w_{m,n}$.

The set of variables is naturally $X = \{x(t_m, t_n), 1 \leq m < n \leq |T|\}$ and all domains are equal to $D = \{(p_{m,n}, w_{m,n}), p_{m,n} \in P, 1 \leq p_{m,n} \leq |P|, w_{m,n} \in W, 1 \leq w_{m,n} \leq |W|\}$; $\forall x \in X$, $D_x = D$. The set C of constraints contains the following three types of constraints:
- Uniqueness of all teams in each week. For each team t_n, $t_n \in T$, $1 \leq n \leq |T|$, we impose the constraint: $\text{WEEK}(t_n) \Leftrightarrow w_{m,n} \neq w_{q,n}$, $\forall (m, q) \in [1; |T|]^2$, $m \neq n$, $q \neq n$ and $m \neq q$;
- No more than two matches for each team in the same period. For each team t_n, $t_n \in T$, $1 \leq n \leq |T|$ and each period, we impose the constraint:
$\text{PERIOD}(t_n) \Leftrightarrow |\{p_{m,n} = p_{q,n}, (m, q) \in [1; |T|]^2, m \neq n, q \neq n \text{ and } m \neq q\}| \leq 1$;
- All matches are different. For each match $< t_m, t_n >$, $(t_m, t_n) \in T^2$, $t_m \neq t_n$, we impose the constraint: $\text{ALLDIFF}(< t_m, t_n >) \Leftrightarrow < t_m, t_n > \neq < t_q, t_r >$, $\forall (t_q, t_r) \in T^2$, $t_q \neq t_r$.
The WEEK and ALLDIFF constraints are always satisfied in our algorithm.

4 Solving the SLSP with Tabu Search

Tabu search is an advanced local search method using general mechanisms and rules as guidelines for smart search. Readers may find formal description of the method in [8]. We now define the components of our tabu algorithm for the SLSP, called TS-SLSP.

4.1 Search Space – Configuration

As represented in Table 1, a configuration is a complete assignment of $D = \{(p_{m, n}, w_{m, n}), p_{m, n} \in P, 1 \leq p_{m, n} \leq |P|, w_{m, n} \in W, 1 \leq w_{m, n} \leq |W|\}$ items to variables of $X = \{x(t_m, t_n), 1 \leq m < n \leq |T|\}$. Thus, a configuration is a $|W| * |P|$ sized table, whose items are integer couples (m, n), $1 \leq m < n \leq |T|$. For $|T| = 40$ teams, this represents a problem with 780 variables and 780 values per variable.

There are $|T| / 2 * (|T| - 1)$ matches to be scheduled. A schedule can be thought of as a permutation of these matches. So, for $|T|$ teams, the search space size is $[|T| / 2 * (|T| - 1)]!$ In other words, the search space size grows as the factorial of the square of $|T| / 2$.

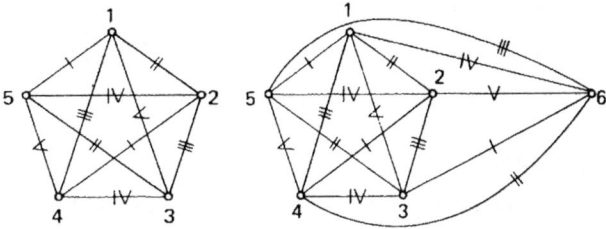

Fig. 1. Constructing initial configuration ($|T| = 6$)

4.2 Initial Configuration

In local search algorithms, the initial configuration specifies where the search begins in the search space. One can start with a random configuration by first creating all the $|T| / 2 * (|T| - 1)$ matches with respect to the ALLDIFF constraint, then randomly assigning a match to each (w, p) couple, $w \in W$, $p \in P$.

We chose another way to build an initial configuration, inspired from [18]: construct a complete graph with the first $|T| - 1$ teams as vertices while placing the vertices in order to form a regular $(|T| - 1)$-gon (edges represent matches). Color the edges around the boundary using a different color for each edge. The remaining edges can be colored by assigning to each one the same color as that used for the boundary edge parallel to it, see left drawing in Fig. 1. At each vertex there will be exactly one color missing and these missing colors are different. The edge of the complete graph incident to the last vertex (i.e., the last team) can be colored using these missing colors, see right drawing in Fig. 1. Finally, fill in week "i", in the initial configuration, with edges colored "i".

This initial configuration has the property of satisfying the WEEK and ALLDIFF constraints. The algorithm will try to satisfy the PERIOD constraint.

4.3 Neighborhood

Let s be a configuration from the search space S. The neighborhood N: S $\rightarrow 2^s$ is an application such as for each $s \in$ S, $s' \in$ N(s) if and only if s and s' only differ by values of a single couple of variables, with at least one in conflict (a variable is said to be conflicting if it is involved in an unsatisfied constraint).

A neighboring configuration of s can be obtained by making a single swap of current values of two arbitrary variables, with at least one conflicting in s. A move from configuration s to a neighboring configuration s' can then be described by a couple $< x(t_m, t_n), x(t_q, t_r) >$, i.e., swapping two matches (see Fig. 2). In addition, swaps are made in the same week to keep the WEEK constraint satisfied.

During the search the neighborhood size always evolves with the number of conflicts.

Weeks								Weeks				
1	2	3	4	5			1	1	2	3	4	5
1,2	2,6	3,4	5,6	2,4	1		1	1,2	2,6	3,4	5,6	**1,5**
4,6	1,3	2,5	1,4	3,6	2	Periods	2	4,6	1,3	2,5	1,4	3,6
3,5	4,5	1,6	2,3	1,5	3		3	3,5	4,5	1,6	2,3	**2,4**

Configuration s   Configuration s'

Fig. 2. Neighborhood illustration (6 teams)

4.4 Evaluation Function

To compare, in terms of quality, two configurations s and s' from S, we define an evaluation function which is an order relation for S.

Let $OccP_s(p, t_n)$ be the occurrence number of team t_n, in period p, in configuration s. The evaluation function f(s) is the total number of excess appearances of all teams in all periods (let us call it $f_P(s)$), i.e., the minimum number of variables to be changed to satisfy the PERIOD constraint:

$$f(s) = f_P(s) = \sum_{p=1}^{|P|} \sum_{n=1}^{|T|} \chi_P(s, n, p),$$

$$\chi_P(s, n, p) = \begin{cases} 0 & \text{if } OccP_s(p, t_n) \leq 2, \\ OccP_s(p, t_n) - 2 & \text{otherwise}. \end{cases}$$

(1)

Solving the SLSP means finding a configuration $s^* \in$ S such as $f(s^*) = 0$.

4.5 Neighborhood Evaluation

The tabu algorithm considers in general at each step the whole neighborhood. So it is imperative to be able to quickly evaluate the cost of neighboring configurations. To do this, we used a technique inspired by [6].

Let δ be a $|X| * |X|$ matrix. Each entry $\delta[x(t_m, t_n), x(t_q, t_r)]$ represents the effect of the chosen move (swapping matches $< t_m, t_n >$ and $< t_q, t_r >$) on the evaluation function.

Thus, the cost of s' ∈ N(s) is immediately obtained by adding the proper entry of δ to f(s) (in $O(1)$). To get the best neighbor, one search only a subset of the δ matrix in $O(|N(s)|)$ time. After a move, the δ matrix is updated in $O(|X|\sqrt{|X|})$ time in the worst case.

4.6 Tabu List Management

A fundamental component of TS is tabu list: a special, short-term memory that maintains a selective history, composed of previously encountered configurations or more generally pertinent attributes of such configurations. The aim of a tabu list is to avoid cycling and go beyond local optima.

Entries in our tabu list are paired matches. Indeed, we make match swaps. After swapping, the matched couple is classified tabu for the next k iterations (k, called tabu tenure, is problem dependent), which means that reverse swap is forbidden during that period.

Tabu tenure plays a very important role for the performance of a tabu algorithm. If overestimated, it will perhaps incorrectly prohibit the visiting of unexplored configurations and the method's capacity to explore the neighborhood will be reduced. If underestimated, the method may get trapped in local optima.

```
Input: |T|; {number of teams}
Output: a valid schedule or "No valid schedule found";
var  f, f*; {evaluation function and its best value
             encountered so far}
     s, s*; {current configuration and the best
             configuration encountered so far}
begin
  initialize the tabu list to empty;
  generate initial configuration s; {section 4.2}
  s* := s;
  f* := f(s);
  while not Stop Condition do
     make a best move m in the same week such that m is
     not tabu or satisfies the aspiration criterion;
     introduce m in the tabu list;
     if (f(s) < f*) then
        s* := s;
        f* := f(s);
     else if f* has not been improved for a sufficient
     time then
        s = s*; {intensification}
        set the tabu list to empty; {diversification}
  if (f* = 0) then {valid schedule found}
     return s*;
  return "No valid schedule found";
end
```

Fig. 3. The TS-SLSP algorithm

After having tried various tabu tenures (randomized and bounded, dynamic, static etc.), we chose a randomized one. For $s \in S$, we define the tabu tenure k_s by: $k_s = \text{rand}(g)$ where $\text{rand}(g)$ is a random integer from $[1; g]$, $g \in [4; 100]$.

To implement the tabu list, we use a $|X| * |X|$ matrix with entries corresponding to moves. Each entry stores the current iteration number plus the tabu tenure. With this data structure, a single comparison with the current iteration number is sufficient to know if a move is tabu.

Nevertheless, note that a tabu move leading to a configuration better than the best configuration found so far is always accepted (aspiration criterion).

4.7 Diversification – Intensification

Intensification and diversification are useful techniques for improving the search power of a tabu algorithm.

Intensification stores interesting properties of the best configurations found so far and to be used later. It may initiate a return to attractive regions to search them more thoroughly. Diversification tries to direct the search to unexplored configurations.

Our TS algorithm includes a single intensification: after a certain number of non-improving iterations, we return to the best configuration found so far. The difficulty then is to formally determine when to go back to this best configuration, because an underestimated interval might inadvertently stop a possibly promising search. Values of this parameter were determined empirically in [30; 1 000].

We make a diversification immediately after an intensification process by removing all tabu statuses to enable previously classified tabu moves.

4.8 General Algorithm

The TS-SLSP algorithm (see Fig. 3) begins with an initial configuration in a search space S built with respect to WEEK and ALLDIFF constraints.

Then it proceeds iteratively to visit a series of locally best configurations following the neighborhood. At each iteration, a best move m (in the same week to keep the WEEK constraint satisfied) is applied to the current configuration, even if it does not improve the current configuration in terms of the value of the evaluation function. Intensification / diversification steps are performed after a sufficient number of non-improving moves.

Stop Condition. The algorithm stops if $f(s) = 0$ (a valid schedule is found) or if a given limit is reached concerning the number of moves.

5 Computational Results

In this section, we compare results of our TS algorithm with those of [11], [15] and [16]. Our tests were carried out on a Sun Sparc Ultra 1 (256 RAM, 143 MHz). TS-SLSP[1] is implemented in C (CC compiler with -O5 option). Tests were carried out

---

[1] For the source code of TS-SLSP, contact the authors.

on problem instances including 6 to 40 teams (only those greater than 14 are shown). The TS-SLSP algorithm was allowed to run until a maximum number of 10 million iterations.

5.1 Comparison Criteria

We used three main criteria to do the comparative study. These criteria are given here in decreasing order of importance:

- Number |T| of teams: one algorithm is said to be more efficient than another one, if it solves the SLSP with a higher |T|;
- Number of moves: the algorithm's effort to find a solution, machine independent;
- Running time: the CPU time spent by an algorithm to carry out a given number of moves, machine dependent.

5.2 Comparative Results

McAloon et al. [11] recall that Integer Linear Programming (ILP) with CPLEX [5] was not able to solve the SLSP for |T| = 14 teams, even when avoiding symmetries. Nevertheless, the method is able to provide solutions for |T| ≤ 12. They also propose a constraint programming algorithm, implemented under ILOG Solver, which solved the |T| = 14 problem in 45 minutes (Ultra Sparc), and was the first result obtained with constraint programming. Finally, with a basic local search algorithm, they solved the |T| = 14 problem in 10 minutes.

Gomes et al. [9] used constraint programming to solve the SLSP up to 18 teams (in more than 48 hours) with a deterministic complete search. By including randomness in their algorithm, they reached the same result more quickly: approximately 22 hours for 18 teams.

In 1998, Régin presented improved results with a more elaborate approach based on constraint propagation [15] (let us call the approach CP1). This approach integrates advanced techniques including symmetries elimination and new powerful filtering algorithms based on arc consistency [3][13][14]. CP1 solves much larger problem instances since it is able to produce valid schedules for |T| = 24 in 12 hours.

In yet another study, by using multiple threads on a 14 processor Sun system, Wetzel and Zabatta [22] obtain better results, since schedules were generated for 26 and 28 teams.

Recently, Régin proposed another improved approach [16] (let us call it CP2) integrating among others, a heuristic based on Euler's theorem for Latin squares. The SLSP is also transformed into an equivalent problem by adding a dummy week in order to quickly deduce some inconsistencies. This later approach gives the best-known results for the SLSP, since it produces solutions for 40 teams in 6 hours.

Table 2 shows the results of TS-SLSP together with those of CP1 and CP2. Columns 2-5 give respectively, for CP1 and CP2, execution times and numbers of backtracks. The last three columns give respectively the success ratio (number of successful runs / total number of runs), mean times and average numbers of moves for successful runs of TS-SLSP.

Table 2. Comparative results of TS-SLSP and well-known methods. A "-" sign means no result is available

|T|	CP1		CP2		TS-SLSP		
	Time	Backtracks	Time (s)	Backtracks	Success Ratio (%)	Time	Moves
16	4.2 s	1 112	0.6	182	100	0.5 s	4 313
18	36 s	8 756	0.9	263	100	0.3 s	2 711
20	< 6 min	72 095	1.2	226	100	23.9 s	149 356
22	10 h	6 172 672	1.2	157	100	34.3 s	163 717
24	12 h	6 391 470	10.5	2 702	90	5 min	1 205 171
26	-	-	26.4	5 683	50	10.7 min	2 219 711
28	-	-	316	32 324	50	12.5 min	2 043 353
30	-	-	138	11 895	10	22 min	3 034 073
32	-	-	-	-	10	49 min	6 387 470
34	-	-	-	-	30	25 min	2 917 279
36	-	-	-	-	10	1.5 h	9 307 524
40	-	-	6 h	2 834 754	10	54.3 s	68 746

From the Table 2, we may make several remarks. First, we observe that TS-SLSP is much more powerful than CP1 both in terms of the size of problem instances solved and computing efforts. Compared with CP2, TS-SLSP also manages to find valid schedules for problem instances going up to 40 teams. We notice however that the resolution becomes much more difficult for |T| > 28. We notice also that the running time of the TS-SLSP algorithm for |T| = 40 is not consistent with other results. This is simply explained by the stochastic nature of the TS-SLSP algorithm. Globally, the results of TS-SLSP are consistent and its running times remain reasonable.

6   Discussions

How to handle the different constraints of a constrained problem is an important issue that must be answered by any neighborhood algorithm. For the SLSP, recall that there are three types of constraints (see section 3.2): WEEK, PERIOD and ALLDIFF.

As we saw previously in section 4, TS-SLSP works with a limited search space whose configurations satisfy already the WEEK and ALLDIFF constraints. The goal of TS-SLSP is then to satisfy the PERIOD constraint.

We also experimented with another constraint handling technique, in which the search space is defined by all the configurations satisfying the ALLDIFF constraint, and the search starts with a randomized initial configuration (see section 4.2). Then the algorithm tries to satisfy the WEEK and PERIOD constraints. In this case, the evaluation function f '(s) used to guide the search is defined by a weighted[2] summation function of the violations of the WEEK constraint $f_w(s)$, formally described by formula

---

[2] It seems that the best weight values are 1 for PERIOD and 2 for WEEK.

2 below[3], and PERIOD constraint $f_P(s)$ (see formula 1 in section 4.4): $\forall\ s \in S, f\ '(s) = 2 * f_W(s) + f_P(s)$.

$$f_W(s) = \sum_{w=1}^{|W|} \sum_{n=1}^{|T|} \chi_W(s, n, w),$$

$$\chi_W(s, n, w) = \begin{cases} 0 & \text{if } OccW_S(w, t_n) \le 1, \\ OccW_S(w, t_n) - 1 & \text{otherwise}. \end{cases} \tag{2}$$

The neighborhood is extended to all possible exchanges of matches (with at least one in conflict) in the same week or in the same period or in different periods and weeks, see Fig. 4.

The tabu tenure is weighted and dynamic: $k\ '_s = \alpha * [f_W(s) + f_P(s)] + rand(g)$, $\alpha \in [0.1; 1]$. Intensification and diversification processes are used in the same way as in TS-SLSP.

		Weeks						
		1	2	3	4	5	6	7
Periods	1	1 vs 3	1 vs 2	5 vs 8	1 vs 5	4 vs 8	2 vs 6	3 vs 5
	2	3 vs 4	2 vs 8	1 vs 4	6 vs 8	2 vs 5	1 vs 7	6 vs 7
	3	5 vs 6	4 vs 6	2 vs 7	4 vs 7	3 vs 7	2 vs 8	1 vs 8
	4	7 vs 8	5 vs 7	3 vs 6	2 vs 3	1 vs 6	4 vs 5	2 vs 4

Fig. 4. Extended neighborhood illustration

Experimentation using this technique produces results similar to those of CP1. Although it fails to solve the |T| > 22 problem, this technique is very fast for 22 teams: it provides solutions in less than 28 minutes (34 146 moves). When only dealing with the PERIOD constraint, TS-SLSP outperforms results of this technique, since it solves the SLSP up to 40 teams. The main difference between the two approaches is the structure of the initial configuration. The technique starting with an extra constraint to verify (the WEEK constraint) must explore a larger search space than the other technique.

7 Conclusion

In this paper we presented a tabu algorithm (TS-SLSP) for the sports league scheduling problem. The algorithm is based on a CSP formulation of the SLSP and includes the following main features:

- a simple swap neighborhood;
- efficient data structures for fast neighborhood search;
- a dynamic tabu tenure;
- a simple intensification-diversification process.

---

[3] $OccW_S(w, t_n)$ is the occurrence number of team t_n on week w in configuration s.

TS-SLSP was tested on problem instances going up to 40 teams. The experimental results show that TS-SLSP largely outperforms some previously developed approaches (ILP [11], basic LS [11] and constraint programming [9][15]). Indeed, while these approaches are limited to instances of 24 teams, TS-SLSP is able to find a schedule for instances going up to 40 teams. This result compares well with the best-known approach, which combines well-elaborated constraint propagation algorithms and a non-trivial formulation of the initial problem [16].

At the same time, the computing times required by TS-SLSP are much greater than those obtained using the most efficient algorithm. Nevertheless, there is certainly plenty of room for improvement in TS-SLSP. One possible enhancement would be to integrate more elaborate intensification and diversification mechanisms, another would be to have a close study of configuration structures in order to devise clever neighborhoods and constraint handling techniques. It would also be interesting to envisage a combination of some advanced constraint propagation techniques with tabu search.

This work underlines once again the importance of efficient data structures for the high performance of a TS algorithm. It also confirms that parameter setting is crucial for obtaining quality solutions.

The TS-SLSP may also be adaptable to other sports scheduling problems, e.g., the minimization of breaks (two consecutive home or away matches) [21] and those studied in [17]. Let us also mention that TS has already proven efficient for solving some sports scheduling problems, with constraints of various types. A hybrid tabu search has been proposed [4] to respond to the National Hockey League's scheduling problem, which underlines the effectiveness of a TS / genetic algorithm combination; the skilful mix of elements of the two methods produces better results than those of separately considered methods. All these studies suggest that a metaheuristic approach (e.g., TS, simulated annealing, genetic hybrid) has good potential for solving various planning and scheduling problems related to sporting events.[4]

Acknowledgments. We would like to thank J.C. Régin for having sent us his latest results and the referees for their relevant remarks.

References

1. Ackley, D.H.: *A Connectionist Machine for Genetic Hillclimbing*. Boston, MA: Kluwer Academic Publishers (1987)
2. Bean, J.C., Birge, J.R.: Reducing Travelling Costs and Player Fatigue in the National Basketball Association. *Interfaces* 10 / 3 (1980) 98–102
3. Bessière, C., Régin, J.C.: Arc Consistency for General Constraint Networks: Preliminary Results. In: *Proceedings of the Fifteenth International Joint Conference on Artificial Intelligence*. Nagoya, Aichi, Japan: International Joint Conference on Artificial Intelligence, Inc. (1997) 398–404
4. Costa, D.: An Evolutionary Tabu Search Algorithm and the NHL Scheduling Problem. *INFOR* 33 / 3 (1995) 161–178

---

[4] Additional studies are reported in [4][19][23].

5. CPLEX Optimization: Using the CPLEX©Callable Library. Tahoe, NV, Inc. (1995)
6. Ferland J.A., Fleurent, C.: Genetic and Hybrid Algorithms for Graph Coloring. *Annals of Operations Research: Metaheuristics in Combinatorial Optimization* 63 / 1. Hammer, P.L. et al. (Eds) (1996) 437–461
7. Ferland J.A., Fleurent, C.: Allocating Games for the NHL Using Integer Programming. *Operations Research* 41 / 4 (1993) 649–654
8. Glover, F., Laguna, M.: *Tabu Search.* Boston, MA: Kluwer Academic Publishers (1997)
9. Gomes C.P., Selman B., Kautz H.A.: Boosting Combinatorial Search Through Randomization. In: *Proceedings of the Fifteenth National Conference on Artificial Intelligence.* AAAI Press / The MIT Press, Madison, WI (1998) 431–437
10. Henz, M.: Scheduling a Major College Basketball Conference – Revisited. To appear in: *Operations Research* 49 / 1 (2001)
11. McAloon, K., Tretkoff, C., Wetzel, G.: Sports League Scheduling. In: *Proceedings of the Third ILOG Optimization Suite International Users' Conference.* Paris, France (1997)
12. Nemhauser, G.L., Trick, M.A.: Scheduling a Major College Basketball Conference. *Operations Research* 46 / 1 (1998) 1–8
13. Régin, J.C.: A Filtering Algorithm for Constraints of Difference in CSPs. In: *Proceedings of the Twelfth National Conference on Artificial Intelligence*, Vol. 1. Seattle, Washington: AAAI Press (1994) 362–367
14. Régin, J.C.: Generalized Arc Consistency for Global Cardinality Constraint. In: *Proceedings of the Thirteenth National Conference on Artificial Intelligence / Eighth Conference on Innovative Applications of Artificial Intelligence*, Vol. 1. Portland, Oregon: AAAI Press (1996) 209–215
15. Régin, J.C.: Modeling and Solving Sports League Scheduling with Constraint Programming. *First Congress of the French Operational Research Society (ROADEF).* Paris, France (1998)
16. Régin, J.C.: Sports Scheduling and Constraint Programming. *INFORMS.* Cincinnati, Ohio (1999)
17. Schaerf, A.: Scheduling Sport Tournaments Using Constraint Logic Programming. *Constraints* 4 / 1 (1999) 43–65
18. Schreuder, J.A.M.: Constructing Timetables for Sport Competitions. *Mathematical Programming Study* 13 (1980) 58–67
19. Terril, B.J., Willis, R.J.: Scheduling the Australian State Cricket Season Using Simulated Annealing. *Journal of the Operational Research Society* 45 / 3 (1994) 276–280
20. Tsang, E.P.K.: Foundations of Constraint Satisfaction. London and San Diego: Academic Press (1993)
21. Werra (de), D.: Geography, Games and Graphs. *Discrete Applied Mathematics* 2 (1980) 327–337
22. Wetzel G., Zabatta F.: CUNY Graduate Center CS Technical Report (1998)
23. Wright, M.: Timetabling County Cricket Fixtures Using a Form of Tabu Search. *Journal of the Operational Research Society* 45 / 7 (1994) 758–770

Appendix:

The table gives a solution to the Sports League Scheduling Problem found by TS-SLSP for 40 teams.

																				Weeks																			
	1	2	3	4	5	6	7	8	9	10	11	12	13	14	15	16	17	18	19	20	21	22	23	24	25	26	27	28	29	30	31	32	33	34	35	36	37	38	39
1	1/2	2/3	3/4	4/5	5/6	6/7	7/8	8/9	9/10	10/11	11/12	32/40	27/39	14/15	15/16	16/17	17/18	18/19	19/20	20/21	21/22	22/23	23/24	24/25	25/13	1/28	27/29	28/30	29/31	30/40	12/33	32/34	33/35	34/36	35/37	36/38	37/39	38/26	14/26
2	3/39	1/4	2/5	3/6	19/31	5/8	6/9	22/34	8/11	9/12	10/13	11/14	12/15	13/16	14/17	15/18	7/28	38/40	18/21	19/31	20/23	21/24	22/25	23/26	24/27	25/28	26/29	27/30	4/16	29/32	30/33	7/19	32/35	33/36	34/37	17/40	36/39	1/37	2/38
3	8/34	5/39	10/36	11/37	3/8	4/9	23/31	6/12	7/17	4/14	9/23	2/20	7/17	12/18	13/19	14/20	15/21	16/26	13/40	1/22	19/26	16/27	21/36	22/29	15/30	24/35	25/32	27/28	28/33	25/18	30/28	10/38	2/39	28/31	38/32	39/37	31/37	5/32	6/37
4	5/37	6/38	7/39	1/8	25/40	3/10	4/11	5/12	6/13	7/14	8/15	9/16	10/17	11/18	12/19	13/20	14/21	15/22	16/23	17/24	18/25	19/26	20/27	21/28	22/29	23/30	24/31	9/40	26/33	27/34	28/35	29/36	30/37	31/38	32/39	1/33	2/34	3/35	4/36
5	9/33	7/37	8/38	9/39	1/10	2/11	3/12	4/13	5/14	6/15	7/16	8/17	30/36	10/19	11/20	12/21	13/22	14/23	15/24	16/25	17/26	18/27	19/28	20/29	6/40	22/31	23/32	24/33	25/34	26/35	27/36	28/37	29/38	30/39	1/31	2/32	18/40	4/34	5/35
6	7/35	8/36	9/37	10/38	11/39	1/12	2/13	3/14	4/15	5/16	6/17	7/18	8/19	9/20	10/21	11/22	12/23	13/24	14/25	15/26	2/40	17/28	18/29	19/30	20/31	21/32	22/33	23/34	24/35	25/36	26/37	27/38	28/39	1/40	16/31	3/32	4/33	5/34	6/34
7	4/38	9/35	1/6	2/7	12/38	13/39	27/40	2/15	3/16	5/13	6/18	7/19	8/16	9/21	10/22	11/23	12/24	13/25	14/22	15/27	20/28	17/25	18/30	19/31	20/32	21/33	26/34	23/31	24/36	29/37	26/34	14/39	32/40	3/37	4/29	1/30	10/35	7/28	8/33
8	18/24	10/34	11/35	12/36	13/37	14/38	24/40	1/16	2/17	30/40	4/19	5/20	33/40	7/22	8/23	33/39	10/31	11/32	12/29	13/34	14/35	6/39	15/36	19/31	3/32	19/35	20/36	21/37	22/38	23/39	9/15	1/25	2/26	3/27	4/28	5/29	15/21	7/31	8/32
9	10/32	11/33	12/34	17/31	14/36	26/40	16/38	21/35	1/18	2/37	3/20	25/39	5/22	2/27	7/24	36/40	9/26	6/31	11/28	12/29	13/30	10/35	15/32	16/33	17/34	14/39	1/15	20/37	21/38	4/18	5/19	2/24	3/25	8/22	9/23	6/28	7/29	8/30	13/27
10	11/31	12/32	13/33	14/34	15/35	16/36	17/37	18/38	19/39	1/20	2/21	3/22	4/23	34/40	6/25	7/26	8/27	9/28	10/29	11/30	12/31	13/32	14/33	15/40	16/35	17/36	18/37	19/38	20/39	1/21	2/22	3/23	4/24	5/25	6/26	7/27	8/28	9/29	10/30
11	12/30	13/31	14/32	15/33	16/34	17/35	18/36	28/40	20/38	21/39	1/22	10/15	3/24	4/25	5/26	6/27	4/31	8/29	9/30	19/32	11/33	12/37	10/35	14/28	23/38	16/39	17/19	18/20	1/21	2/40	3/23	13/24	5/25	6/26	7/27	8/23	9/33	2/34	11/29
12	13/29	14/30	15/31	17/32	18/33	19/34	20/35	21/36	22/37	2/38	31/40	1/24	2/25	3/26	4/27	5/28	28/29	7/30	39/40	9/33	10/34	11/35	12/36	13/37	14/38	15/39	16/17	1/18	18/19	3/20	13/21	5/22	6/23	7/24	8/25	9/26	10/27	11/28	12/28
13	14/28	15/29	16/30	13/35	18/32	19/33	20/34	17/39	22/36	2/19	24/38	4/21	1/26	6/23	3/28	4/29	5/30	10/27	7/32	8/33	9/34	14/31	11/36	12/37	13/38	18/35	8/40	2/16	3/17	22/39	1/23	6/20	7/21	22/40	9/27	10/24	11/25	12/26	9/31
14	15/27	22/40	17/29	18/30	4/7	20/32	21/33	19/37	23/35	24/36	25/37	26/38	13/14	1/28	2/29	3/30	16/19	5/32	6/33	7/34	8/35	4/36	11/40	12/38	26/39	2/27	3/14	28/15	5/31	6/17	31/18	8/34	9/20	10/21	11/22	12/23	13/24	11/25	13/39
15	16/26	17/27	18/28	24/40	20/30	21/31	22/32	23/33	24/34	25/35	26/36	27/37	28/29	1/30	2/31	3/32	4/33	5/34	6/35	7/36	8/37	9/38	10/39	1/11	12/13	3/14	4/15	5/16	6/17	7/18	8/19	9/20	10/21	11/22	12/23	13/40	14/41	19/25	15/25
16	17/25	18/26	19/27	20/28	21/29	22/30	5/10	24/32	25/33	26/34	27/35	28/36	29/37	30/38	31/39	1/32	2/33	3/34	4/35	18/23	6/37	3/40	8/39	1/9	2/10	7/40	4/12	5/13	6/14	7/15	8/16	9/17	31/36	11/18	12/19	20/21	14/22	15/23	16/24
17	6/36	19/25	23/40	21/27	22/28	23/29	15/39	25/31	26/32	27/33	28/34	29/35	18/37	37/38	24/34	1/35	2/36	3/37	4/38	30/7	3/8	5/30	2/10	11/40	12/11	13/12	14/13	4/14	39/18	16/17	10/18	11/19	12/20	13/21	14/22	20/23	3/16	22/40	16/40
18	19/23	20/24	21/25	22/26	23/27	24/28	25/29	26/30	27/31	28/32	29/33	30/34	31/35	32/36	35/40	34/37	37/1	1/2	2/3	3/4	4/5	2/37	3/38	4/39	5/6	6/7	7/8	8/9	9/10	10/11	11/12	12/13	13/14	14/15	15/16	16/17	17/18	19/20	18/21
19	20/22	21/23	22/24	23/25	24/26	25/27	26/28	27/29	28/30	29/31	30/32	31/33	32/34	33/35	34/36	35/37	36/38	37/39	38/38	39/39	1/3	2/4	3/5	4/6	5/7	6/8	7/9	8/10	9/40	10/40	11/13	12/14	13/15	14/16	15/17	16/17	17/18	18/19	19/21
20	21/41	16/28	20/26	19/29	2/9	15/37	1/14	7/10	29/40	3/18	23/39	12/13	6/21	5/24	33/37	8/25	35/39	17/20	8/31	5/36	16/27	7/38	13/34	15/34	18/33	3/11	19/36	25/32	9/11	10/12	31/32	4/26	1/30	2/38	27/30	26/24	6/24	14/23	17/23

Lagrange Multipliers for Local Search on Planning Graphs

Alfonso Gerevini and Ivan Serina

Dipartimento di Elettronica per l'Automazione
Università di Brescia, via Branze 38
25123 Brescia, Italy
{serina,gerevini}@ing.unibs.it

Abstract. GPG is a planner based on planning graphs that combines local search and backtracking techniques for solving both plan-generation and plan-adaptation tasks. The space of the local search is formed by particular subgraphs of a planning graph representing partial plans. The operators for moving from one search state to the next one are graph modification operations corresponding to adding (deleting) actions to (from) a partial plan. GPG can use different types of heuristics based on a parametrized cost function, where the parameters weight different types of constraint violation that are present in the current subgraph. A drawback of this method is that the performance is sensitive to the static values assigned to these parameters.

In this paper we propose a refined version of the local search heuristics of GPG using a cost function with dynamic parameters. In particular, the cost of the constraint violations are dynamically evaluated using Lagrange multipliers. As the experimental results show, the use of these multipliers gives two important improvements to our local search. First, the revised cost function is more informative and can discriminate more accurately the elements in the neighborhood. As a consequence, the new cost function can give better performances. Secondly, the performance of the search does not depend anymore on the values of the parameters that in the previous version need to be tuned by hand before the search.

1 Introduction

Domain-independent planning is a very hard search problem. The large majority of the search techniques that have been proposed for domain-independent planning relies on systematic methods that in principle can examine the complete search space (e.g., [1,10,11]). However, despite their theoretical completeness, in practice these algorithms are incomplete because for many planning problems the search space is too large to be (even partially) explored, and so a plan cannot be found in reasonable time (if one exists).

Recently, alternative approaches using local search have been proposed (e.g., [2,3,5,8,9]). These methods are formally incomplete, but in practice they can be more efficient than systematic methods when the planning problem is solvable.

A. Nareyek (Ed.): Local Search for Planning and Scheduling, LNAI 2148, pp. 37–54, 2001.
© Springer-Verlag Berlin Heidelberg 2001

In [3] we introduced a framework for local search in the context of the "planning through planning graph analysis" approach [1]. We formulate the problems of generating and adapting a plan as search problems, where the elements of the search space are particular subgraphs of the planning graph representing partial plans. The operators for moving from one search state to the next one are particular graph modification operations corresponding to adding (deleting) some actions to (from) the current partial plan.

The general search scheme consists of an iterative improvement process which, starting from an initial subgraph of the planning graph, greedily improves the "quality" of the current plan according to a cost function. Such function measures the cost of the graph modifications that are possible at any step of the search. The cost of a modification is estimated in terms of the constraint violations that a modification introduces or eliminates, where a constraint violation is either an unsatisfied action precondition or an exclusion relation between actions. A final state of the search process is any subgraph representing a valid complete plan.

This local search framework has been implemented in a system called GPG, which can use different types of heuristics to guide the search. These heuristics can be easily implemented by appropriately setting the values of certain parameters of the cost function weighting the different types of constraint violations that can be introduced or removed by a graph modification.

Unfortunately, preliminary experimental results indicated that the performance of this original technique can be highly sensitive to the static values assigned to these parameters, and in this paper we give additional experimental results showing that indeed this is the case. In order to cope with this drawback, we propose a refined version of the local search heuristics of GPG using a cost function with *dynamic* parameters. Our technique is similar to the use of discrete Lagrange multipliers recently proposed in [13,14,15] for solving SAT problems. The general idea is that a set of constraint violations is associated with a dynamic multiplier estimating the cost of solving them. If the value of the multiplier is high, then the corresponding set of constraint violations is considered "difficult", while if the value is low it is considered "easy". The multipliers have all the same default initial value, which is dynamically updated during the search each time a local minimum is reached.

As the experimental results show, the use of these multipliers in the cost function guiding the search gives two important improvements to our local search. First, the revised cost function is more informative and can discriminate more accurately the elements in the neighborhood. As a consequence, the new cost function can give better performances. Secondly, the performance of the search does not depend anymore on the values of the parameters in the cost function, that in the previous version need to be tuned by hand before the search.

The rest of this paper is organized as follows. In Section 2 we briefly illustrate the planning graph data structure; in Section 3 we introduce our general local search scheme; in Section 4 we illustrate the use of Lagrange multipliers in the cost function; in Section 5 we present our experimental results; finally, in Section 6 we give our conclusions.

2   Planning Graphs

In this section, we briefly review the approach to planning based on planning graphs [1]. Given a planning problem, first we construct a planning graph for it, and then we search for a particular subgraph of the planning graph representing a valid plan for the problem.

A planning graph is a directed acyclic levelled graph with two kinds of nodes and three kinds of edges. The levels alternate between a fact level, containing fact nodes, and an action level containing action nodes. An action node al level t represents an action that can be planned at time t. A fact node represents a proposition corresponding to a precondition of one or more operators instantiated at time step t (actions at time step t), or to an effect of one or more actions at time step $t - 1$. The fact nodes of level 0 represent the positive facts of the initial state of the planning problem.[1] The last level is a proposition level containing the fact nodes corresponding to the goals of the planning problem.

In the following, we indicate with $[u]$ the proposition (action) represented by the fact node (action node) u. The edges in a planning graph connect action nodes and fact nodes. In particular, an action node a of level i is connected by:

- *precondition edges* from the fact nodes of level i representing the preconditions of $[a]$;
- *add-edges* to the fact nodes of level $i + 1$ representing the positive effects of $[a]$;
- *delete-edges* to the fact nodes of level $i + 1$ representing the negative effects of $[a]$;

Two action nodes of a certain level are *mutually exclusive* if no valid plan can contain both the corresponding actions. Similarly, two fact nodes are mutually exclusive if no valid plan can make both the corresponding propositions true. There are two cases in which two action nodes a and b are marked as mutually exclusive in the graph [1]: one of the actions deletes a precondition or add-effect of the other (*interference*); a precondition node of a and a precondition node of b are marked as mutually exclusive (*competing needs*).

Two proposition nodes p and q in a proposition level are marked as exclusive if all the ways of making proposition $[p]$ true are exclusive with all ways of making $[q]$ true. Specifically, they are marked as exclusive if each action node a having an add-edge to p is marked as exclusive of each action node b having an add-edge to q. When two facts or actions are marked as mutually exclusive, we say that there is a *mutex relation* between them.

An action node a of level i can be in a "valid subgraph" of the planning graph (a subgraph representing a valid plan) only if all its precondition nodes

---

[1] Planning graphs adopt the Closed World Assumption.

are *supported*, and a is not involved in any mutual exclusion relation with other action nodes of the subgraph. We say that a fact node q of level i representing a proposition $[q]$ is supported in a subgraph $\mathcal{G}'$ of a planning graph $\mathcal{G}$ if either

- in $\mathcal{G}'$ there is an action node at level $i - 1$ representing an action with (positive) effect $[q]$, or
- $i = 0$ (i.e., $[q]$ is a proposition of the initial state).

Given a planning problem $\mathcal{P}$, the corresponding planning graph $\mathcal{G}$ is incrementally constructed level by level starting from level 0 using a polynomial algorithm [1]. The last level of the graph is the first propositional level where the goal nodes are present, and there is no mutex relation between them. (In some cases, when the problem is not solvable, the algorithm identifies that there is no level satisfying these conditions, and hence it detects that the problem is unsolvable before starting to search).

Once a planning graph has been constructed, we search for a subgraph $\mathcal{G}'$ of $\mathcal{G}$ that is a *solution* (valid plan) for $\mathcal{P}$. I.e., we search for a subgraph such that

(1) all the precondition nodes of the actions in $\mathcal{G}'$ are supported,
(2) every goal node is supported,
(3) there are no mutual exclusion relations between action nodes of $\mathcal{G}'$.

If the search of a solution fails, the planning graph can be extended by adding extra levels, and a new search is performed on the resulting graph. The reasons why a search phase can fail depends on the kind of search that is performed. In the case of systematic search, as in IPP [10], the search fails because it detects that the problem is unsolvable given the current number of levels in the graph; while in the case of a local search, as in GPG, the search fails because a certain search limit has been exceeded.

In order to simplify the notation in the rest of the paper, we will consider the goal facts in the last level of the graph as precondition facts of a special additional action node, *End*, which has no effect and which is the last action in any planning graph.

Figure 1 gives an example of a planning graph for a simple problem in the Rocket domain, that was originally given in [4]. The action load(b,R,S) is mutually exclusive with move(R,S,D) because move(R,S,D) deletes a precondition of load(b,R,S). The fact in(b,R) and at(R,S) are mutually exclusive because all the ways of generating the first (i.e., load(b,R,S)) are exclusive of all the ways of generating the second (i.e., move(R,S,D)).

3   Local Search on Planning Graphs

In this section we review the general local search process of GPG that we proposed in [3].

The local search of GPG for a planning graph $\mathcal{G}$ of a given problem $\mathcal{P}$ is a process that, starting from an initial subgraph $\mathcal{G}'$ of $\mathcal{G}$ (a partial plan for $\mathcal{P}$),

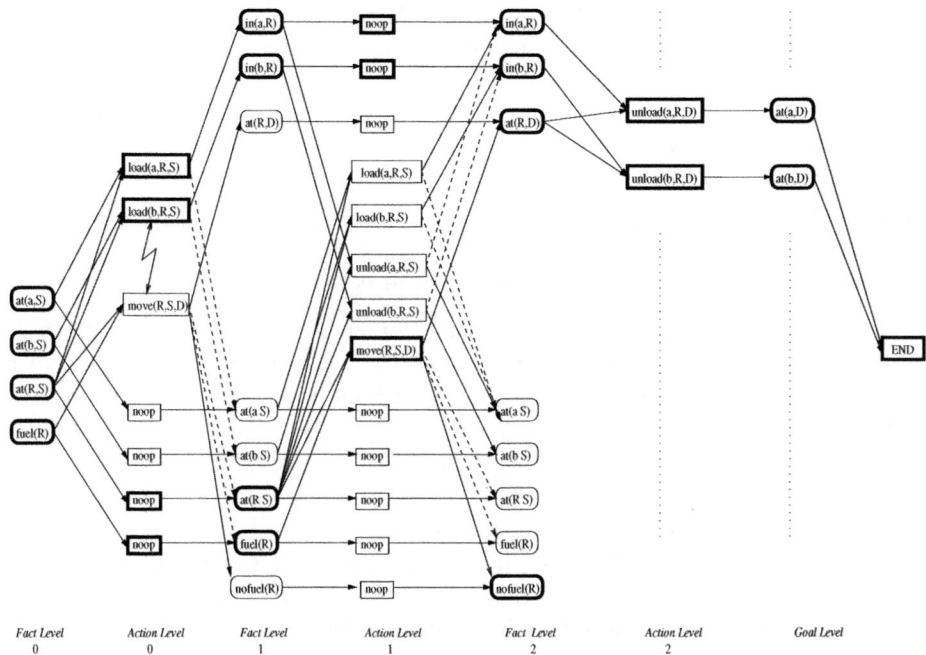

Fig. 1. A portion of the planning graph for a simple problem in the Rocket domain [1] with one rocket R, two objects a and b, a start location S and destination D. Delete edges (negative effects) are represented by dashed lines, add edges and precondition edges by solid lines. The solution plan is represented by facts and actions with bold boxes.

transforms $\mathcal{G}'$ into a solution of $\mathcal{P}$ through the iterative application of some graph modifications improving the quality of the current partial plan. Each modification is either an extension of the subgraph to include a new action node of $\mathcal{G}$, or a reduction of the subgraph to remove an action node (and the relevant edges).

Adding an action node to the subgraph corresponds to adding an action to the partial plan represented by the subgraph (analogously for removing an action node). At any step of the search process the set of actions that can be added or removed is determined by the **constraint violations** that are present in the current subgraph of $\mathcal{G}$. Such violations correspond to

- mutually exclusive relations involving action nodes in the current subgraph;
- unsupported facts, which are either preconditions of actions in the current partial plan, or goal nodes in the last level of the graph (i.e., precondition nodes of the special action End).

More precisely, the search space is formed by the set **A** of all the *action subgraphs* of the planning graph $\mathcal{G}$.

An **action subgraph** $\mathcal{A} \in \mathbf{A}$ of a planning graph $\mathcal{G}$ is a subgraph of $\mathcal{G}$ containing the final action *End* and such that, if a is an action node of $\mathcal{G}$ in $\mathcal{A}$, then also the fact nodes of $\mathcal{G}$ corresponding to the preconditions and positive effects of $[a]$ are in $\mathcal{A}$, together with the edges of $\mathcal{G}$ connecting them to a.

A **solution subgraph** of $\mathcal{G}$ (a final state of the search problem) is an action subgraph $\mathcal{A}_s$ containing the goal nodes of $\mathcal{G}$ and such that

(1) all the fact nodes corresponding to preconditions of actions in $\mathcal{A}_s$ are supported;
(2) there is no mutex relation between action nodes.

The general scheme for searching a solution graph consists of two main steps. The first step is an initialization of the search in which we construct an initial action subgraph. The second step is a local search process in the space of all the action subgraphs, starting from the initial action subgraph. We can generate an initial action subgraph in several ways. Three possibilities that we have implemented are:

— a randomly generated action-subgraph;
— an action-subgraph where all precondition facts and goal facts are supported (but in which there may be some violated mutually exclusive relations);
— an action-subgraph obtained from an existing plan given in input to the process.

These kinds of initialization can be performed in linear time with respect to the size of $\mathcal{G}$. The third initialization method is particularly useful when the local search is used for solving plan adaptation problems (further details on the initialization step can be found in [5]).

Once we have computed an initial subgraph, the search phase is performed in the following way. A constraint violation in the current subgraph is randomly chosen. If it is an unsupported fact node, then in order to resolve (eliminate) this constraint violation, we can either add an action node that supports it, or we can remove an action node which is connected to that fact node by a precondition edge. If the constraint chosen is an exclusion relation, then we can remove one of the action nodes of the exclusion relation. Note that the elimination of an action node can remove several constraint violations (i.e., all those corresponding to the set of exclusion relations involving the action node eliminated). On the other hand, the addition of an action node can introduce several new constraint violations. Also, when we add (remove) an action node to satisfy a constraint, we also add (remove) all the edges connecting the action node with the corresponding precondition and effect nodes in the planning graph – this ensures that each change to the current action subgraph is another action subgraph.

More formally, we can define the neighborhood of an element of the search space in the following way: given a planning graph $\mathcal{G}$, an action subgraph $\mathcal{A}$ of $\mathcal{G}$ and a constraint violation s in $\mathcal{A}$, the **neighborhood** $N(s, \mathcal{A})$ of s in $\mathcal{A}$ is the set of action subgraphs obtained from $\mathcal{A}$ by applying a graph modification that resolves s.

At each step of the local search the elements in the neighborhood are weighted according to a certain cost function, and an element with the minimum cost is then chosen as the next subgraph (search state).

Given a planning problem $\mathcal{P}$, a planning graph $\mathcal{G}$ of $\mathcal{P}$ and an action subgraph $\mathcal{A}$ of $\mathcal{G}$, the **general cost function** of $\mathcal{A}$ is a function $F : \mathbf{A} \to N$ such that:

$$F(\mathcal{A}) = \sum_{a \in \mathcal{G}} mutex(a, \mathcal{A}) + precond(a, \mathcal{A})$$

where:

$$mutex(a, \mathcal{A}) = \begin{cases} 0 & \text{if } a \notin \mathcal{A} \\ me(a, \mathcal{A}) & \text{if } a \in \mathcal{A} \end{cases}$$

$$precond(a, \mathcal{A}) = \begin{cases} 0 & \text{if } a \notin \mathcal{A} \\ pre(a, \mathcal{A}) & \text{if } a \in \mathcal{A} \end{cases}$$

and $me(a, \mathcal{A})$ is the number of action nodes in $\mathcal{A}$ which are mutually exclusive with a; $pre(a, \mathcal{A})$ is the number of precondition facts of a which are not supported in $\mathcal{A}$.

It is easy to see that, when the search reaches a subgraph for which the value of F is zero, the planning problem has been solved (the current subgraph is a solution subgraph). Using F we can also formulate the local search process on $\mathcal{G}$ as the following constrained optimization problem:

$min_{\mathcal{A} \in \mathbf{A}} F(\mathcal{A})$

subject to
$mutex(a, \mathcal{A}) = 0$ for all $a \in \mathcal{G}$
$precond(a, \mathcal{A}) = 0$ for all $a \in \mathcal{G}$

4   Heuristics

4.1   Some Basic Heuristics

The simple use of a general cost function to guide the local search has the drawback that it can lead to local minima from which the search can not escape. For this reason, instead of using F, GPG uses a parametrized **action cost function** C which is more flexible than F, and allows to specify different types of heuristics. This function specifies the cost of inserting (C^i) and of removing (C^r) an action $[a]$ in the partial plan represented by the current action subgraph $\mathcal{A}$. The action cost function is defined in the following way:

$$C([a], \mathcal{A}) = \begin{cases} C([a], \mathcal{A})^i = \alpha^i \cdot pre(a, \mathcal{A}) + \beta^i \cdot me(a, \mathcal{A}) + \gamma^i \cdot unsup(a, \mathcal{A}) \\ C([a], \mathcal{A})^r = \alpha^r \cdot pre(a, \mathcal{A}) + \beta^r \cdot me(a, \mathcal{A}) + \gamma^r \cdot sup(a, \mathcal{A}), \end{cases}$$

where $me(a, \mathcal{A})$ and $pre(a, \mathcal{A})$ are defined as in the general cost function, $unsup(a, \mathcal{A})$ is the number of unsupported precondition facts or goals in $\mathcal{A}$ that become supported by adding a to $\mathcal{A}$, and $sup(a, \mathcal{A})$ is the number of supported precondition facts or goals in $\mathcal{A}$ that become unsupported by removing a from $\mathcal{A}$.

By appropriately setting the values of the coefficients α, β and γ we can implement different heuristic methods aimed at making the search less influenced by local minima because "less committed" to follow the gradient of the general cost function. Their values are set before searching and have to satisfy the following constraints:

- $\alpha^i, \beta^i > 0, \gamma^i \leq 0$ in C^i
- $\alpha^r, \beta^r \leq 0, \gamma^r > 0$ in C^r.

Note that the positive coefficients of C (α^i, β^i and γ^r) determine an increment of C which is related to an increment of the number of constraint violations. Analogously, the non-positive coefficients of C (α^r, β^r and γ^i) determines a decrement of C which is related to a decrement of the number of constraint violations.

In [3], we have proposed three basic heuristics: Walkplan, Tabuplan and T-Walkplan. Walkplan is similar to the heuristic used by Walksat [12,9], and can be implemented by setting $\alpha^i, \beta^i > 0, \gamma^i = 0$ and $\alpha^r, \beta^r = 0, \gamma^r > 0$. I.e., the action cost function of Walkplan is

$$C([a], \mathcal{A}) = \begin{cases} C([a], \mathcal{A})^i = \alpha^i \cdot pre(a, \mathcal{A}) + \beta^i \cdot me(a, \mathcal{A}) \\ \\ C([a], \mathcal{A})^r = \gamma^r \cdot sup(a, \mathcal{A}). \end{cases}$$

In Walkplan, the best element in the neighborhood is the subgraph introducing the fewest new constraint violations (i.e., it does not consider the constraint violations of the current subgraph that are resolved). In addition, Walkplan uses a "noise" parameter p. Given an action subgraph $\mathcal{A}$ and a (randomly chosen) constraint violation, if there is a modification that does not introduce new constraint violations, then the corresponding action subgraph in $N(s, \mathcal{A})$ is chosen as the next action subgraph. Otherwise, with probability p one of the subgraphs in $N(s, \mathcal{A})$ is chosen randomly, and with probability $1 - p$ the next action subgraph is chosen according to minimum value of the action cost function.

Tabuplan is a tabu search [6,7] in which the last k actions inserted or removed in the action subgraph are stored in a tabu list of length k. At each step of the search, from the current action subgraph we choose as next subgraph the best subgraph in $N(s, \mathcal{A})$ which can be generated by adding or removing an action that is *not* in the tabu list. The best subgraph in $N(s, \mathcal{A})$ is chosen using C with the coefficients set to values satisfying $\alpha^i, \beta^i > 0, \gamma^i < 0$ and $\alpha^r, \beta^r < 0, \gamma^r > 0$. I.e., this heuristic is more greedy than Walkplan, in the sense it considers not only the new constraint violations that are introduced by graph modification, but also those that are eliminated.

Finally, T-Walkplan is a combination of Walkplan and Tabuplan in which a tabu list is used simply for *increasing the cost* of a certain graph modifications, instead of preventing them as in Tabuplan. More precisely, when we evaluate the cost of adding or removing an action $[a]$ which is in the tabu list, the action cost $C([a], \pi)$ is incremented of $\delta \cdot (k - j)$, where δ is a small quantity (e.g., 0.1), k is the length of the tabu list, and j is the position of $[a]$ in the tabu list. (Further details on the basic heuristics implemented in GPG can be found in [3,5].)

4.2 Dynamic Heuristics Based on Lagrange Multipliers

Unfortunately, experimental results revealed that the values of the α, β and γ coefficients in C can significantly affect the efficiency of the search. Since their value is static, the optimal performance can be obtained only when their values are appropriately "tuned" before search. Moreover, the best parameter setting can be different for different planning domains (or even for different problems in the same domain). In order to cope with this drawback, we have revised the cost functions by weighting their terms with *dynamic* coefficients similar to the discrete *Lagrange multipliers* that have successfully used for solving SAT problems [13,15].

The use of these multipliers gives two important improvements to our local search. First, the revised cost function is more informative, and can discriminate more accurately the elements in the neighborhood. Secondly, the new cost function does not depend on any static coefficient (the α, β and γ coefficients can be omitted, and the initial default value of all the new multipliers is relatively important).[2]

The general idea is that each constraint violation is associated with a dynamic multiplier that weights it. If the value of the multiplier is high, then the corresponding constraint violation is considered "difficult", while if the value is low it is considered "easy". Using these multipliers, the cost of a subgraph will be estimated not only in terms of the number of constraint violations that are present in it, but also in terms of an estimation of the difficulty to solve the particular constraint violations that are present.

Since the number of constraint violations that can arise during the search can be very high, instead of maintaining a distinct multiplier for each of them, our technique assigns a multiplier to a *set* of constraint violations associated with the same action. Specifically, for each action a, we have a multiplier λ_m^a for all the mutex relations involving a, and a multiplier λ_p^a for all the preconditions of a. Intuitively, these multipliers estimate how difficult is to globally satisfy all the preconditions of an action, and to globally avoid to have that a is mutually

---

[2] However, when both positive and negative terms are considered as in Tabuplan, using a static parameter for weighting the relative impact on the cost function could be useful to impose a particular degree of greediness to the search. The higher is the weight of the negative terms, the more greedy is the search.

exclusive with another action in the action subgraph. As we will show, this estimation is based on the search already performed up to the current action subgraph. The multipliers can be used in the general cost function to weight the unsupported preconditions or goals, and mutex relations that are present in the current action subgraph. In particular, the function F can be revised to the following function:

$$F_\lambda(\mathcal{A}) = \sum_{a \in \mathcal{G}} (\lambda_m^a \cdot mutex(a, \mathcal{A}) + \lambda_p^a \cdot precond(a, \mathcal{A}))$$

Similarly, we can refine the cost $C([a], \mathcal{A})^i$ of inserting the action node $[a]$ and the cost $C([a], \mathcal{A})^r$ of removing $[a]$ as follow:

$$C_\lambda([a], \mathcal{A})^i = \lambda_p^a \cdot pre(a, \mathcal{A}) + \lambda_m^a \cdot me(a, \mathcal{A}) - \Delta_a^+$$

$$C_\lambda([a], \mathcal{A})^r = -\lambda_p^a \cdot pre(a, \mathcal{A}) - \lambda_m^a \cdot me(a, \mathcal{A}) + \Delta_a^-$$

where Δ_a^+ is the sum of the unsupported preconditions of the actions a_j in $\mathcal{A}$ that become supported by adding a, weighted by $\lambda_p^{a_j}$; while Δ_a^- is the sum of the supported preconditions that become unsupported by removing a, weighted by $\lambda_p^{a_j}$. More formally, the Δ-terms are defined in the following way:

$$\Delta_a^+ = \sum_{a_j \in \mathcal{A}} \lambda_p^{a_j} \cdot (pre(a_j, \mathcal{A}) - pre(a_j, \mathcal{A} + \{a\})).$$

$$\Delta_a^- = \sum_{a_j \in \mathcal{A} - \{a\}} \lambda_p^{a_j} \cdot (pre(a_j, \mathcal{A} - \{a\}) - pre(a_j, \mathcal{A})),$$

For example, the action cost function for Walkplan becomes

$$C_\lambda([a], \mathcal{A})^i = \lambda_p^a \cdot pre(a, \mathcal{A}) + \lambda_m^a \cdot me(a, \mathcal{A})$$

$$C_\lambda([a], \mathcal{A})^r = \Delta_a^-.$$

The multipliers have all the same default initial value, which is dynamically updated during the search. Intuitively, we want to increase the multipliers associated with actions introducing constraint violations that turn out to be difficult to solve, while we want to decrease the multipliers associated with actions whose constraint violations tend to be easier. These will lead to prefer subgraphs in the neighborhood that appear to be globally easier so solve (i.e., that are closer to a solution graph).

The update of the multipliers is performed whenever the local search reaches a local minimum (i.e., when there is no element in the neighborhood that has a C-cost equal to, or less than zero). The values of the multipliers of the actions that are responsible of the constraint violations present in the current action subgraph are then increased, while the multipliers of the actions that do not determine any constraint violation are decreased. For example, if a is in the current subgraph and is involved in a mutex relation with another action in the subgraph, then λ_m^a is increased; similarly, if a has a precondition that is not supported, λ_p^a is increased. In order to prevent the unbounded increase and decrease of the multipliers, we impose a maximum and a minimum value to them (λ_{max} and λ_{min} respectively). More precisely, when the search reaches a local minimum we update the λ-multiplers in the following way:

$$\lambda_p^a = \begin{cases} min(\lambda_{max}, \lambda_p^a + \delta^+) & \text{if } precond(a, \mathcal{A}) \neq 0 \\ \\ max(\lambda_{min}, \lambda_p^a - \delta^-) & \text{if } precond(a, \mathcal{A}) = 0 \end{cases}$$

$$\lambda_m^a = \begin{cases} min(\lambda_{max}, \lambda_m^a + \delta^+) & \text{if } mutex(a, \mathcal{A}) \neq 0 \\ \\ max(\lambda_{min}, \lambda_m^a - \delta^-) & \text{if } mutex(a, \mathcal{A}) = 0 \end{cases}$$

where δ^+ is a positive small quantity controlling how fast the multipliers increase their value, and δ^- is a positive small quantity controlling how fast the multipliers decrease their value.[3]

5   Experimental Results

In this section, we give some experimental results concerning the performance of GPG for a set of planning problems, some of which are very hard to solve for both Graphplan and IPP, two planner using backtrack search on planning graphs.[4] In particular, we have considered problems in the following known benchmark domains: Rocket (rocket_a), Logistics (logistics_a, logistics_b and logistics_c), Tsp (tsp_11), Fridgeworld (f2), Sokoban (sk1) and Gripper (gripper_8).

The experiments were aimed at testing the behavior of the local search process with and without the Lagrange multipliers. Here we focus on the results for Walkplan (other experiments aimed at testing Tabuplan and T-Walkplan

---

[3] In order to make more accurate the increment of λ_p^a, in our implementation δ^+ is weighted by taking into account the proportion k of unsupported preconditions of a with respect to the total number of unsupported preconditions in the action subgraph (i.e, λ_p^a is increase by $k \cdot \delta^+$). Similarly, when we increase λ_m^a, δ^+ is weighted by the proportion of mutex involving a with respect to the total number of mutex in the subgraph.

[4] GPG is available by inquiry to the authors.

SECONDS (logarithmic scale)

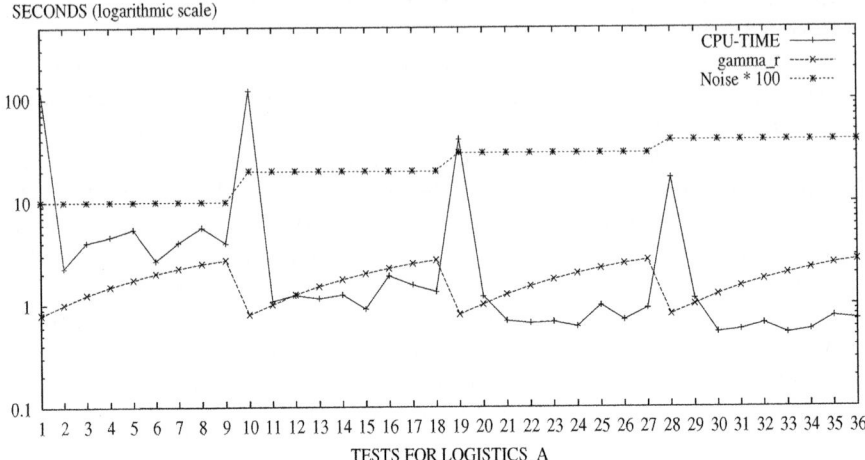

Fig. 2. Performance of Walkplan without Lagrange multipliers for logistics_a.

SECONDS (logarithmic scale)

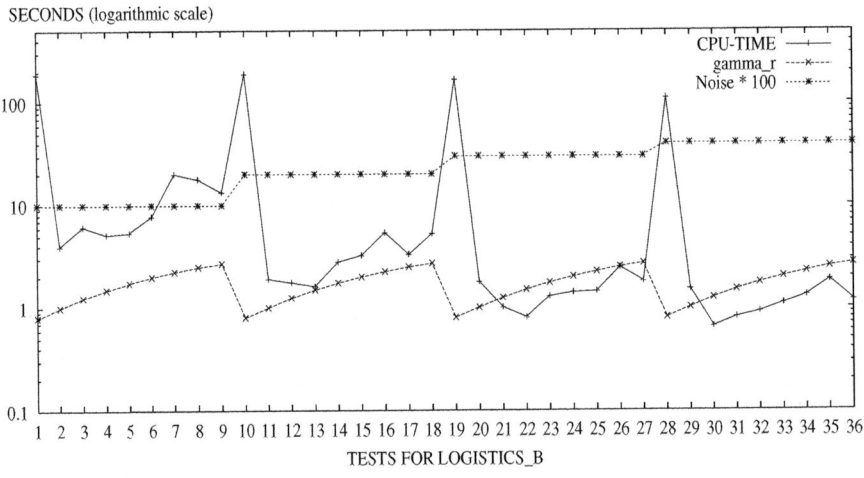

Fig. 3. Performance of Walkplan without Lagrange multipliers for logistics_b.

are in progress). The tests were conducted on a PC Pentium II 400 MHz with 256 Mbytes of RAM. The planning graph was constructed with the minimum number of levels that is sufficient to solve the problem.

Figures 2, 3, 4, 5, 6, 7, 8 and 9 plot the CPU-time in logarithmic scale required by Walkplan using the original cost function without the Lagrange multipliers. Each figure refers to a different benchmark problem. On the x-axis we have the tests corresponding to different values of the parameters γ^r and p (the noise);

SECONDS (logarithmic scale)

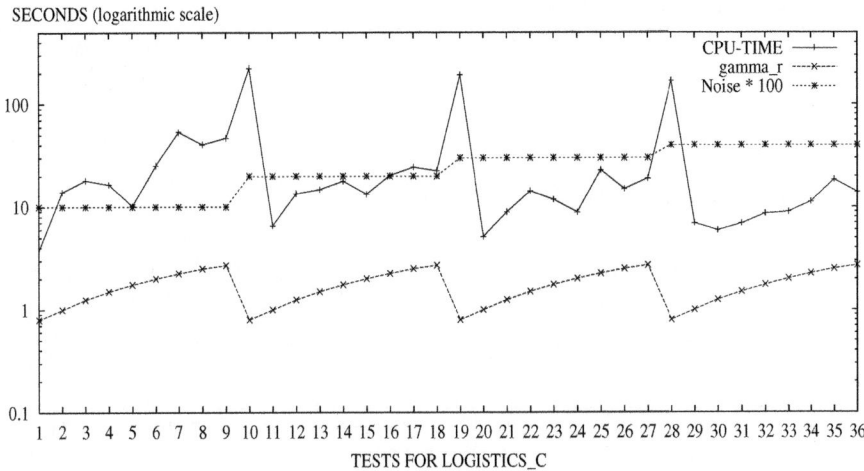

TESTS FOR LOGISTICS_C

Fig. 4. Performance of Walkplan without Lagrange multipliers for logistics_c.

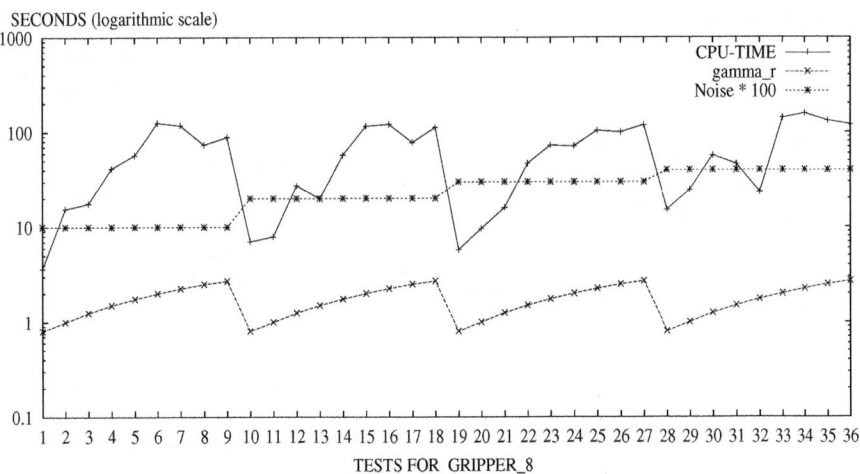

TESTS FOR GRIPPER_8

Fig. 5. Performance of Walkplan without Lagrange multipliers for gripper_8.

γ^r was varied between 0.8 and 2.7, p was varied between 0.1 to 0.4.[5] The values of these parameters are plotted as different curves. In particular, for each of the four noise values considered, nine different values of γ^r were used. Each CPU-

---

[5] The values for α^i and β^i were constantly kept equal to 1. The reason for this is that we had experimentally observed that better performance can be obtained when these parameters are set to the same value, and that Walkplan is mainly influenced by the ratio of the values of α^i and β^i (i.e., the parameters associated with the introduction of new constraint violations) to the value of γ^r (i.e., the parameter associated with the removal of existing constraint violations).

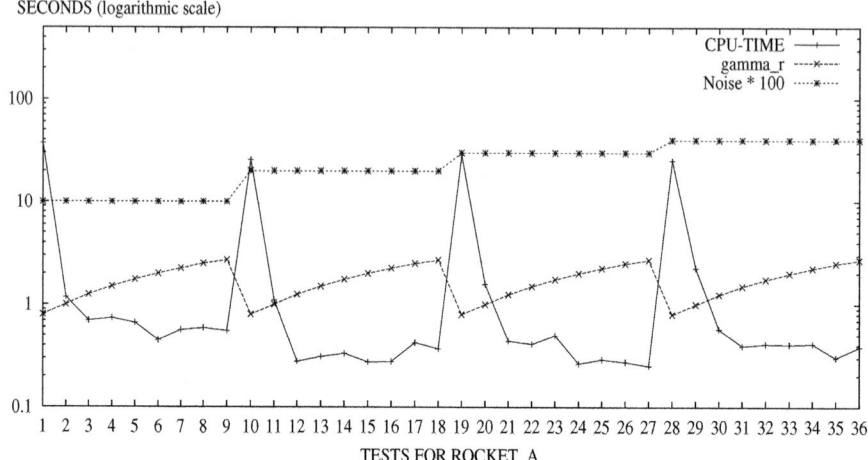

Fig. 6. Performance of Walkplan without Lagrange multipliers for rocket_a.

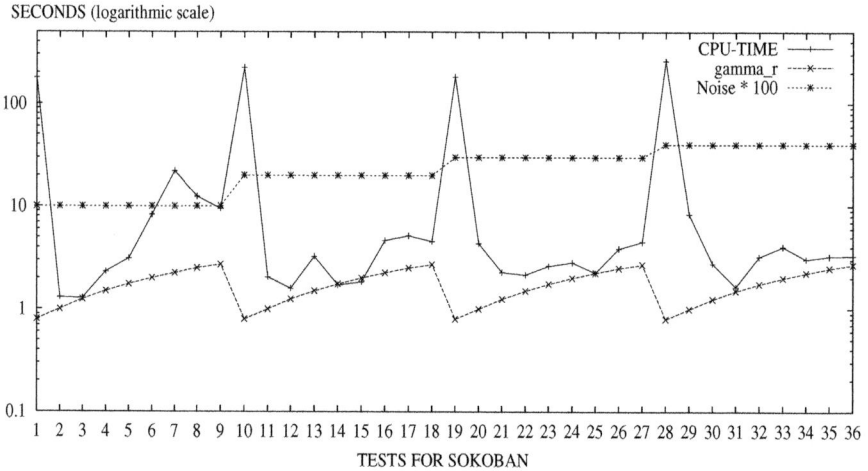

Fig. 7. Performance of Walkplan without Lagrange multipliers for sk1 in the sokoban domain.

time plotted in the figures is an average over 20 runs; each run consists of at most ten tries; each try consists of an initialization phase (the generation of an initial action subgraph), followed by a search phase where the parameters of the heuristic were set as indicated.

The results of these figures show that the performance of Walkplan is highly sensitive to the values of the parameters of the cost function (in logistics_b, for example, when the noise is set to 0.1 the minimum CPU-time required was 3.9 seconds and the maximum 188.8).

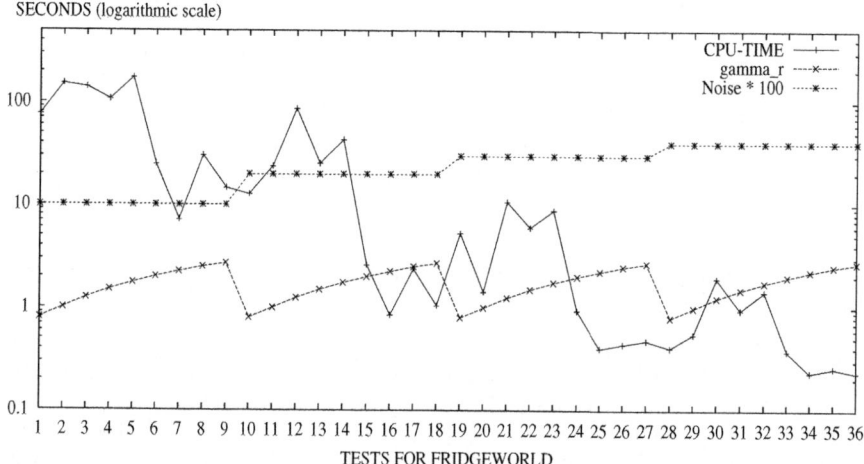

Fig. 8. Performance of Walkplan without Lagrange multipliers for f2 in the Fridge-world domain.

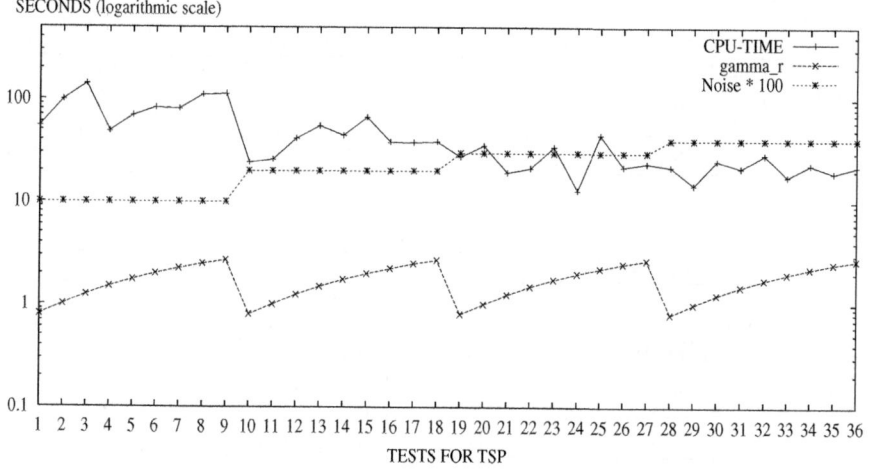

Fig. 9. Performance of Walkplan without Lagrange multipliers for tsp_11 in the tsp domain.

Figure 10 gives the average CPU-times of Walkplan with the new action cost function using the Lagrange multipliers (C_λ). Each CPU-time is an average over 20 runs. Each table in the figure refers to a different benchmark problem and includes the minimum/maximum CPU-times required by Walkplan without La-grange multipliers (C) in the previous experiments.[6] The results for C_λ given in Figure 10 show that the performance of the local search using the new technique

---

[6] Note that these results are plotted in Figures 2–9 and are averages over 20 runs.

Logistics_a			
Noise	C			C_λ
	min	mean	max	mean
0.1	2.27	18.5	134.4	1.19
0.2	0.9	14.5	120.1	0.61
0.3	0.6	5.23	40.6	0.48
0.4	0.52	2.5	17.2	0.49

Logistics_b			
Noise	C			C_λ
	min	mean	max	mean
0.1	3.9	29.8	188.8	1.35
0.2	1.62	23.9	190	0.92
0.3	0.80	19.8	166	0.88
0.4	0.64	13.24	109.8	0.92

Logistics_c			
Noise	C			C_λ
	min	mean	max	mean
0.1	3.9	25.4	53.5	2.52
0.2	6.55	39.6	223.3	3.1
0.3	5.1	33.0	192	1.2
0.4	5.9	27.7	168.8	1.8

Gripper_8			
Noise	C			C_λ
	min	mean	max	mean
0.1	3.6	59.94	88.3	1.97
0.2	7.0	60.56	120.4	2.6
0.3	5.7	60.46	103.7	19.5
0.4	23.3	79.3	156	62.9

Rocket_a			
Noise	C			C_λ
	min	mean	max	mean
0.1	0.44	4.74	37.2	0.59
0.2	0.27	3.21	25.6	1.01
0.3	0.25	3.62	28.5	0.71
0.4	0.3	3.37	25.2	1.14

Sokoban_1			
Noise	C			C_λ
	min	mean	max	mean
0.1	1.2	26.47	178.2	1.19
0.2	1.5	27.51	222.8	1.12
0.3	2.1	23.0	182.5	1.8
0.4	1.6	32.1	259	7.0

Fridgeworld_2			
Noise	C			C_λ
	min	mean	max	mean
0.1	7.1	80.18	172.3	5.8
0.2	0.8	21.95	86.0	6.1
0.3	0.4	3.83	10.7	1.5
0.4	0.23	0.71	1.9	0.36

Tsp_11			
Noise	C			C_λ
	min	mean	max	mean
0.1	48.7	88.5	140.1	33.2
0.2	14.1	41.37	66.4	19.2
0.3	12.9	26.88	45.1	17.0
0.4	14.6	21.06	29.2	12.0

Fig. 10. CPU-time (seconds) required by Walkplan with (C_λ) and without (C) Lagrange multipliers for solving some benchmark problems.

is close to, or even better than the best performance that can be obtained using the original technique with static parameters.

6   Conclusions

Local search for domain independent planning has recently emerged as a promising alternative method to traditional backtracking techniques. GPG is a recent system for plan generation and adaptation that combines local search and backtracking techniques [3,5]. In this paper we have focused on the local search of GPG, proposing a new dynamic cost function for guiding the search. The resulting search technique has two main advantages over the previous technique relying on appropriately setting some static parameters of the cost function. First, the original parameters of the cost function have been replaced with dynamic parameters (Lagrange multipliers), deriving a function that can give a more accurate evaluation of the elements in the neighborhood. Secondly, since these parameters are automatically tuned during the search, the performance of the local search is much less sensitive to their initial value, which can be the same default value for all the planning problems and domains.

The results of several experimental tests using Walkplan show that the performance of the new local search technique of GPG is close to, or better than the best results we can obtain when the parameters of the original cost function are tuned by hand before the search.

Current work includes further experiments to test the performance of Tabuplan and T-Walkplan using dynamic multipliers in the action cost function. Preliminary results suggest that the use of Lagrange multipliers can improve also the performance of these heuristics.

References

1. A. Blum and M.L. Furst. Fast planning through planning graph analysis. *Artificial Intelligence*, 90:281–300, 1997.
2. B. Bonet and H. Geffner. Planning as heuristic search. *Artificial Intelligence*, 2000. (Special Issue on Heuristic Search) Forthcoming.
3. A. Gerevini and I. Serina. Fast planning through greedy action graphs. In *Proceedings of the 16th National Conference of the American Association for Artificial Intelligence (AAAI-99)*, pages 503–510, Orlando, Florida, July 1999. AAAI Press / MIT Press.
4. A. Gerevini and I. Serina. Plan adaptation through planning graph analysis. In *AI*IA: Advances in Artificial Intelligence. Lecture Notes in Artificial Intelligence 1792*, pages 356–367. Springer-Verlag, 1999.
5. A. Gerevini and I. Serina. Fast plan adaptation through planning graphs: Local and systematic search techniques. In *Proceedings of the 5th International Conference on Artificial Intelligence Planning Systems (AIPS-00)*, Austin, Colorado, 2000. AAAI Press / MIT Press.
6. F. Glover and M. Laguna. Tabu search. In C. R. Reeves, editor, *Modern heuristics for combinatorial problems*. Blackwell Scientific, Oxford, GB, 1993.

7. F. Glover, E. Taillard, and D. de Werra. A user's guide to tabu search. *Annals of Operation Research*, 41:3–28, 1993.
8. J. Hoffmann. A heuristic for domain independent planning and its use in an enforced hill-climbing algorithm. Technical Report 133, Institut für Informatik, Universität Freiburg, 2000.
9. H.A. Kautz and B. Selman. Pushing the envelope: Planning, propositional logic, and stochastic search. In *Proceedings of the Thirteenth National Conference of the American Association for Artificial Intelligence (AAAI-96)*, Portland, OR, 1996.
10. J. Koehler, B. Nebel, Hoffmann J., and Y. Dimopoulos. Extending planning graphs to an ADL subset. In *Fourth European Conference on Planning (ECP'97)*, pages 273–285. Springer Verlag, 1997.
11. J.S. Penberthy and D.S. Weld. UCPOP: A sound, complete, partial order planner for ADL. In Bernard Nebel, Charles Rich, and William Swartout, editors, *Proceedings of the Third International Conference on Principles of Knowledge Representation and Reasoning (KR'92)*, pages 103–114, Boston, MA, 1992. Morgan Kaufmann.
12. B. Selman, H.A. Kautz, and B. Cohen. Noise strategies for improving local search. In *Proceedings of the Twelfth National Conference of the American Association for Artificial Intelligence (AAAI-94)*, pages 337–343, Seattle, WA, 1994. Morgan Kaufmann.
13. Y. Shang and B. W. Wah. A discrete Lagrangian-based global-search method for solving satisfiability problems. *Journal of Global Optimization*, 12 no. 1:61–99, 1998.
14. B. W. Wah and Z. Wu. The theory of discrete Lagrange multipliers for nonlinear discrete optimization. In *Proc. Principles and Practice of Constraint Programming*, pages 28–42. Springer-Verlag, Oct. 1999.
15. Zhe Wu and Benjamin W. Wah. An efficient global-search strategy in discrete Lagrangian methods for solving hard satisfiability problems. In *Proceedings of the 17th National Conference of the American Association for Artificial Intelligence (AAAI-00)*, pages 310–315. AAAI Press / MIT Press, July 2000.

Beyond the Plan-Length Criterion

Alexander Nareyek

GMD FIRST, Kekuléstr. 7
D - 12489 Berlin, Germany
alex@ai-center.com
http://www.ai-center.com/home/alex/

Abstract. Resources such as food, hit points or magical power are standard features of today's computer games. Computer-guided characters often have goals that are related to these resources and must take this into account in planning their behavior. In the field of artificial intelligence, resource-based action planning systems have recently begun to mushroom. However, these systems still tend to neglect resources. Conventional plan length is the primary optimization goal, resources being of only secondary importance. This paper illustrates the problem and gives an overview of the EXCALIBUR agent's planning system, which overcomes these deficiencies by applying an extended constraint programming framework to planning. It allows the search space to be explored without the need to focus on plan length, makes it possible to optimize other criteria like resource-related properties and promotes the inclusion of domain-dependent knowledge to guide and accelerate the search for a plan.

1 Introduction

A key factor in today's computer games is the computation of an appropriate behavior of computer-guided characters/mobiles/items (henceforth called *agents*). Reactive capabilities can be easily integrated by means of simple hard-wired if-then rules. Higher-level reasoning is hardly ever implemented, however. This is not because such methods are not needed. The reason is that there are hardly any techniques available because the field of artificial intelligence (AI) has focused mainly on simplified problems and algorithms with provable properties, such as completeness and termination. However, problems of optimizing an agent's behavior with respect to resource-related properties, such as the maximization of an agent's strength, belong to the complexity class of undecidable problems and are hardly ever addressed.

The AI field that deals with determining the right actions to pursue an agent's goals is *AI planning*[1]. The basic planning problem is given by an initial world description, a partial description of the goal world, and a set of actions/operators that map a partial world description to another partial world description. A

---

[1] Please note that the use of the term *planning* in AI is different from what people in the operations research (OR) community would expect (e.g., *scheduling*).

A. Nareyek (Ed.): Local Search for Planning and Scheduling, LNAI 2148, pp. 55–78, 2001.
© Springer-Verlag Berlin Heidelberg 2001

solution is a sequence of actions leading from the initial world description to the goal world description and is called a *plan*. The problem can be enriched by including further aspects, like temporal or uncertainty issues, or by requiring the optimization of certain properties.

Nearly all real-world (virtual-world) problems involve numerical resources, which restrict a planning system's decision options and may be subject to optimization criteria. In recent years, new planning systems have been developed that are able to handle resources. However, in most cases the resources are only used as auxiliary conditions to guarantee a valid plan. Even in most approaches that include an optimization of resource-related properties, the plan length still serves as the primary optimization criterion.

Sections 2 and 3 introduce the Orc Quest example and use it to show the inappropriateness of today's planning systems for resource-optimization problems. Section 4 presents an overview of the EXCALIBUR agent's planning system, which tackles these problems and explains the system's approach using the Orc Quest example. Key techniques are the use of an extended constraint programming framework and the application of local search guided by domain knowledge. Related work and conclusion are given in Section 5.

2   The Orc Quest Example

Imagine you are a computer-guided Orc in a computer game[2]. Your interests are somewhat different from those of your philistine Orc friends. To get them a little more acquainted with the fine arts, you decided to surprise them with a ballet performance. As a highlight, five of those delicate humans are to play supporting roles. However, the uncivilized humans do not appreciate the charm of Orc-produced art and are extremely reluctant to participate. This makes it rather painful to catch a human and force him to take part in the performance, and much more painful to catch a whole group of them. On the other hand, going for a group saves time. This can be formalized as follows:

STATEVARS: Duration, Pain, Performers $\in \mathbb{N}_0$

INIT: Duration = 0, Pain = 0, Performers = 0
GOAL: Performers $\geq$ 5

ACTION catch_only_one:
 Duration += 2, Pain += 1, Performers += 1

ACTION catch_a_group:
 Duration += 5, Pain += 4, Performers += 3

---

[2] For those who have not heard of Orcs before: "Orcs ... are the most famous of Tolkien's creatures. ... Orcs tend to be short, squat and bow-legged, with long arms, dark faces, squinty eyes, and long fangs. ... Orcs hate all things of beauty and love to kill and destroy." [Mattsson 2000]

2.1   The Application of Planning Systems

Being an Orc, you lack sufficient brain power to solve the problem. You therefore apply a conventional AI planning system and come up with the following plan: catch_a_group & catch_a_group, which yields a plan duration of **10 hours**, an **8 on the pain scale** and **6 performers**. The plan attains your goal of catching at least 5 performers, but the other results look to you to be capable of improvement.

You realize that you have forgotten to apply an optimization criterion. You thus repeat the planning process, applying a state-of-the-art resource-based planning system that you request to minimize your pain, hoping dearly that none of your Orc friends will realize what a coward you are.

GOAL: Performers $\geq$ 5, MIN(Pain)

Strangely enough, the resource-based planning system delivers the same plan as the conventional planning system.

2.2   The Real Optimum

Shortly afterward, the first group of humans is caught and you decide to get them to verify the planning system's results. The three gawky simpletons believe your promise to release them if they help you. After a while, they present you with a plan of five sequential catch_only_one actions, which yields a plan duration of **10 hours**, a **5 on the pain scale** and **5 performers**. This answer worries you as it involves much less pain than the solution found by the planning system employed to minimize your pain. Is human AI technology merely a way of undermining the Orcs' preordained dominance in the fine arts?

2.3   A Question of Relevance

You call the humans to account for the behavior of the planning system. They reply that most resource-based planning systems consider only a limited number of actions, increasing this number only if they are unable to produce a *correct plan*, the primary optimization goal being *plan length*. If it is possible to produce a correct plan with a certain number of actions, optimization of other goal properties can begin and plans with a larger number of actions are no longer considered. For your planning problem, the plan catch_a_group & catch_a_group is the only correct plan with a *minimal number of two actions*, and thus also the optimum with respect to any secondary optimization criterion.

Taking the number of actions or the plan length as the primary optimization criterion seems a very academic approach to you and you wonder if it has any relevance. You notice that the humans are becoming increasingly uncomfortable when they start arguing that optimization of the plan's duration is what is needed in most cases. And the duration would be the same as the number of actions. You no longer trust these humans and demand that they draw a picture of the complete search space. The result does not improve the humans' situation

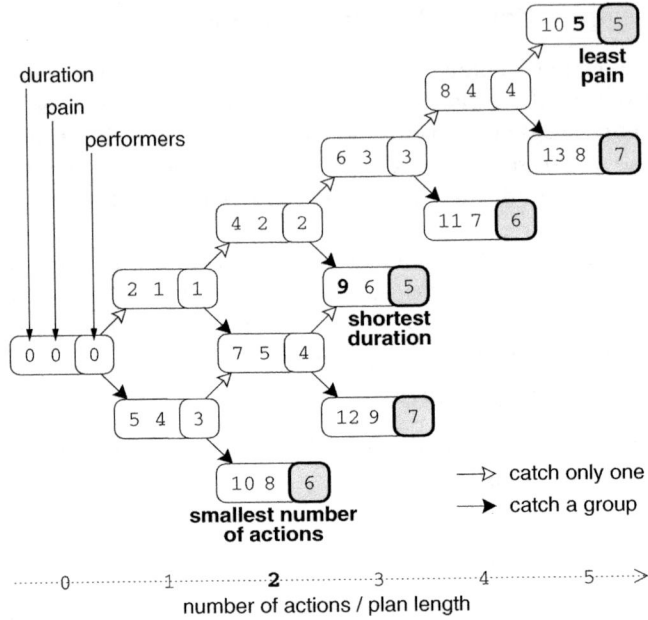

Fig. 1. The Search Space of the Orc Quest Example

because the plan with the shortest duration is different from the one involving the minimal number of actions (see Figure 1).

Luckily for the group of humans, your goal is a plan involving the minimal amount of pain, and you reluctantly release the humans as promised because the plan involving the minimal amount of pain does not entail catching a whole group of humans. Somehow you feel dissatisfied at not having broken your promises.

3   The Nearby Wizard

A little bit later, it occurs to you that there is a wizard living nearby who can prepare a magic potion that lessens the pain by one on the pain scale. The drawback here is that the wizard is also very fond of humans because he is in continual need of test persons to improve his miserable skills in casting teleportation spells. Usually, then, he wants some humans in exchange for his magic potion, which he calls "ethanol".

You visit the wizard and explain your problem to him. He scratches his head, walks around for a while deep in thought and finally says: "All right, this is what we'll do: I'll give you 11 phials of potion for every 10 humans that you promise to bring to me." You are not really sure how many phials you should take and formalize this for your planning system as follows:

```
ACTION deliver_humans:
  Duration += 1, Pain -= 11, Performers -= 10
```

3.1   Plan-Length Bounds

The optimization problem without the `deliver_humans` action could have been optimally solved by calculating an upper bound for the plan length and starting the planning system with the requirement to consider only plans with a length of this bound. A `no_action` action could be introduced to allow for plans that are shorter than the length allowed by the bound. The bound is easy to calculate because a plan's `Performers` and `Pain` increase strictly monotonically with the plan length, and after an application of 5 actions the goal condition for the `Performers` is met anyway, and the `Pain` can only get worse for longer plans.

However, this simple way of calculating an upper bound is no longer applicable with an action like `deliver_humans` that destroys the monotony. You think about the new problem for a bit, but it is far too complex for an Orc. Lacking a method to calculate the upper bound, you decide to run the planning system with some random bounds for the plan length. Starting with a bound of 6, the plan that consists of 5 times `catch_only_one` is returned. With a length of 7, you still get this solution. You decide to raise the bound to 15, and ... wait. After a long period, the solution is confirmed again. It seems to be the optimum, but you test a bound of 30 just to be sure, and ... wait. After a while, the planning system returns a different solution than before. Unfortunately, this solution is beyond your comprehension and you are gradually overcome by the unpleasant feeling that the suggested plan "out of memory" is just another feature of the system designed to mislead helpless Orcs.

The wizard is quite astonished when you tell him that his offer would not help you to improve your plan. You offer to demonstrate this with your planning system, but he wants to try his own. Shortly afterward, he presents you with a plan that entails no pain for you. The plan has a plan length of 60, and you wonder how his system was able to return a solution that quickly.

3.2   A Question of Efficiency

"The wrong approach taken by your planning system," the wizard starts to explain, "is due to its focusing on qualitative problems, like the ancient blocks-world domain. The only feature distinguishing solutions was the plan length, and so everyone wanted to optimize this. A systematic exploration of the search space along an increasing plan length is an obvious way of tackling this problem, and such approaches are still at the heart of today's planning systems, even if the problems now involve optimization criteria that do not change strictly monotonically with the plan length."

In response to your questioningly look, he continues: "Your planning system has a very strange way of exploring the search space." He draws a small figure on the ground (see Figure 2).

"The plan-length optimization of your system," the wizard explains, "tries to verify that there is no valid solution for a certain plan length, and has to perform a complete search for this length before being able to increase the plan length. And even if you instruct it to begin with a certain length bound, it constructs

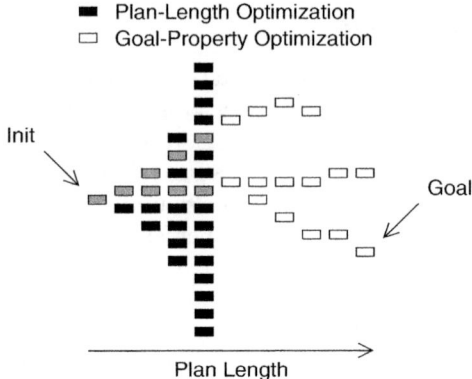

Fig. 2. Search-Space Exploration

the structures for the whole search space for this length, which is a pretty useless approach for non-toy problems.

As you have already realized, the plan-length criterion is a poor guide for the search because it has absolutely no relevance, so the search is rather like a random search. It is much more convenient to explore the search space without being bound to the plan length, as my planning system does. This allows us to perform a goal-property-driven search by integrating domain knowledge. Searching in an infinite search space means relying on powerful heuristics instead of being tied to a method of exploration that is determined by an irrelevant criterion."

3.3   Complexity

The wizard still sees a questioning look on the face in front of him, and he heaves a sigh: "Are you wondering about the infinite search space? Our planning problem is *undecidable* because it involves numbers with unpredictable bounds — you can read up on this in [Erol, Nau & Subrahmanian 1991]."

Suddenly, while casting a light-the-candle spell the wizard's apprentice vanishes in a big ball of fire. He had always been eager to try out this spell, but was warned by the wizard about how dangerous it was. Now the apprentice had been eavesdropping while you were talking with the wizard and had concluded that the optimization of his life's duration is also an undecidable problem, and thus not worth considering.

The wizard shakes his head sadly: "Hardcore theorists! If a problem is undecidable, they hate to tackle it." Just to be sure that you will not fall victim to one of the wizard's teleporation spells, you quickly put in: "I'm not one of them!" "No," the wizard smiles, "Orcs don't tend to be theorists anyway. But perhaps this neglect of undecidable problems is the reason why for so long no one dared to tackle problems other than plan-length optimization."

3.4   Decision-Theoretic Planning

The wizard invites you to dinner, and you gladly accept. A little while later, you are sitting in a comfortable armchair with a jar of the wizard's ethanol potion in your hands. The time has come to ask where he got his planning system from. The wizard's potion seems to have a paralyzing effect on your tongue, but finally you manage to put your question in reasonably comprehensible terms.

The wizard takes a while to answer and gives you an earnest look. Then he raises his eyebrows meaningfully: "Have you heard of EXCALIBUR?" You are seized with a shivering fit because EXCALIBUR is a popular name for swords used by computer players' avatars that generally enjoy tyrannizing harmless Orcs.

"No, no," the wizard says appeasingly, "not the sword!" His voice takes on a vibrant tone: "Did you ever consider the possibility that you are part of a computer program and that your behavior is largely guided by a planning system?" He pauses briefly to let you reflect on all this question's implications. "Well," he continues, casting a thunderbolt to conjure up a more threatening atmosphere, "what I mean is the EXCALIBUR Agent's Planning System!" He lowers his voice: "We are the *agents*. You understand?"

This sounds like one of those popular conspiracy theories. You are, however, not particularly well versed in philosophy and already have difficulty explaining why there are now two wizards and why all of the equipment in the room is swaying curiously. "Please put off describing the system for now," you say, putting an end to this rather dubious discussion. "If resources are such a vital component, why aren't there other planning systems like the EXCALIBUR agent's planning system?"

"Well," the wizards say, "I must admit that there have, of course, also been other approaches at trying to focus on optimizing things other than plan length. These are mostly subsumed under the term *decision-theoretic planning*. But most of the research in this area aims to optimize probabilistic issues for reasoning under conditions of uncertainty (see [Boutilier, Dean & Hanks 1999] for an overview). Although goals like maximizing probability to attain a goal can also be interpreted as resource-related optimization (the probability being the resource to be maximized), probability is usually considered to have only monotonical features, i.e., a plan cannot have a higher probability of attaining the goal than a subplan that also attains the goal. Your problem involving an action like using a magic potion has no counterpart here.

But there are also resource-oriented planning systems that are able to handle more complex problems, e.g., Williamson and Hanks' PYRRHUS planning system [Williamson & Hanks 1994] and Ephrati, Pollack and Milshtein's A*-based approach [Ephrati, Pollack & Milshtein 1996]. The difference between these and the EXCALIBUR agent's planning system is the conceptual approach."

"Sorry, did you say A^*? It's funny that you should mention this term because it sometimes comes mysteriously to mind — usually when I think about how to get to a certain place. Do you know what it means?"

"Well, all it means is that the EXCALIBUR agent's planning system has specialized heuristics for specific subtasks. Don't worry about it! But — A* brings me directly to the point. The resource-oriented planning systems mentioned above construct plans in a stepwise manner, e.g., action by action, and if they fail, they go back to an earlier partial plan and try to elaborate this. In order to choose a partial plan for an elaboration — called *refinement* — an upper bound of the quality is computed for each partial plan. The plan that yields the highest upper bound is chosen for elaboration. Sounds like a great concept, doesn't?"

"Probably not — since you ask!" you reply and applaud loudly, impressed by your own cleverness. The wizards continue: "Er, you're right, the concept is great for things like path finding, for which it is easy to determine bounds because the actions can only add values. But for complex planning problems it is not normally possible to calculate bounds on the quality value for a partial plan. We spoke about the problems with bounds before. Does a single action like building a trap for someone puts bounds on the total fun you will have?"

"Definitely not! Things can actually get worse!" you answer bad-temperedly, "If I don't take precautions, the person trapped may be a player's avatar, and then the world is usually reset to a former state only to build a trap for the person who will build the trap."

"Exactly," the wizards continue, "The EXCALIBUR agent's planning system takes a different approach. It iteratively repairs *complete grounded plans* for which a concrete quality can be determined. The repair methods can exploit the quality information to improve the plan, e.g., by replacing a set of actions that cause a loss in quality. This technique, based on local-search methods, is not so – let's say *unfocused* – with respect to the plan quality, and there is practically no restriction on the kind of objective function to be optimized (see also [ILOG, Inc. 2000]). The ASPEN system [Chien et al. 2000, Rabideau et al. 1999] is very similar to EXCALIBUR in this respect. To sum up, *decision-theoretic planning* is a very broad term, which, of course, also encompasses the EXCALIBUR agent's planning system."

4   The EXCALIBUR Agent's Planning System

"Let me give you an overview of the system", the wizards say. "Please consult the references that I'll give in the following for a more detailed description of the individual techniques.

4.1   Model Structures

In EXCALIBUR, a planning problem is formulated as a constraint satisfaction problem (CSP). A constraint is a test that checks a specific relation between certain variables. For example, the constraint EQUALS can check the equality between two variables. If all of a problem's constraints are fulfilled (or *satisfied* or *consistent*, terms also used here), the variable's values constitute a valid solution to the problem. A CSP specifies which constraints and variables exist and how the variables are related by which constraints. Where the planning *model* is mentioned in the following, this refers to the CSP used to specify the planning problem.

The planning model focuses on resources. A resource is a temporal projection of a specific property's state, which may be subject to constraints such as preconditions and changes. Numerical as well as symbolic properties are uniformly treated as resources. For example, a battery's POWER is a resource, but the PAIN you experience and a DOOR are, too:

POWER is $[\ 0:\ t \in [0..5]\ ,\ 10 - 0.75 \times t:\ t \in [6..13]\ ,\ 0:\ t \in [14..\infty[\ ]$,

PAIN is $[\ 0:\ t \in [0..1]\ ,\ 1:\ t \in [2..\infty[\ ]$,

DOOR is $[$ OPEN: $t \in [0..45]$, CLOSED: $t \in [46..60]$, UNKNOWN: $t \in [61..\infty[\ ].$"

"Why do you call them *resources* if they are merely state variables that can also have a numerical domain?" you ask. "The term *resource*," the wizards explain, "is a term commonly used in the constraint programming (CP) and OR community. If you prefer the terms *state variable* or *fluent*, then use these instead. For the planning model, we use the term *resources* because of the system's close relation to applications for resource allocation/optimization.

Figure 3 shows an example that informally illustrates the planning model's basic elements. Figure 4 gives an overview of the formal CSP representation; it does not show the CSP itself, but it does indicate possible relations between constraint types and variables for the planning CSP. The CSP itself can include multiple instances of the constraints and variables shown. Only some of the constraints are discussed here. For a more detailed description, please refer to [Nareyek 2000].

Oops, sorry, I forgot to tell you what the elements of Figure 4 mean. A variable is depicted by a circle, a constraint by a rectangle. Arrows are used to connect variables to constraints. The role/position of a variable within a constraint can be expressed by an arrow's label and direction.

In addition, there are special types of constraints — so-called *object constraints* — that do not check the variables' values, but provide structural context information. For example, it must be known which two *Begin* and *End* variables together form an ACTION TASK. Otherwise, an ACTION RESOURCE CONSTRAINT designed to check if the connected ACTION TASKs overlap might combine *Begin* and *End* variables of different ACTION TASKs for the check. Object constraints are represented by a rectangle with a dashed outline.

An action (e.g., eating a peanut) consists of a set of different preconditions that must be satisfied (e.g., that the agent has an peanut), operations that the agent has to perform (e.g., the mouth's operations) and resulting state changes (e.g., the agent's hunger being satisfied). These elements are represented by *tasks*, i.e., there are PRECONDITION TASKs for precondition tests, ACTION TASKs for operations and STATE TASKs for state changes. The tasks are collections of variables, e.g., temporal variables that determine the beginning and end of the tasks, or variables that specify how a state's properties are changed by the action. For example, an action `catch_a_group` includes a STATE TASK that has

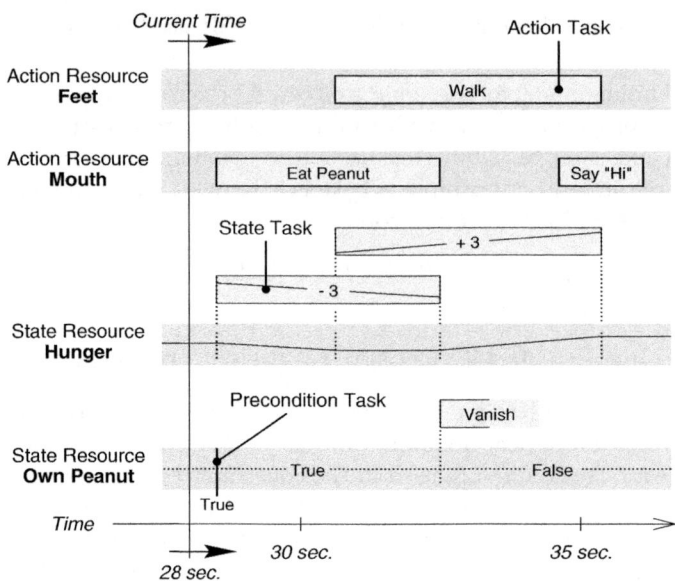

Fig. 3. Visualization of the Basic Plan Elements

a *Contribution* variable with a value of 4, which represents the resulting increase in pain.

ACTION RESOURCE CONSTRAINTs ensure that the agent has enough capacity to carry out the operations that are needed to execute the plan's actions. The ACTION RESOURCE CONSTRAINT correlates ACTION TASKs and ensures that the tasks do not overlap. For your problem, we need only a general ORC ACTION RESOURCE CONSTRAINT, as there is no further refinement regarding your hands, feet, mouth or whatever. The ORC ACTION RESOURCE CONSTRAINT can be linked to ACTION TASKs with a duration of 2 hours (if the task's *Operation* variable has the value catch_only_one), 5 hours (for a value of catch_a_group) and 1 hour (for a value of deliver_humans).

A STATE RESOURCE CONSTRAINT temporally projects the state of a specific environment property and checks if the plan's related preconditions are satisfied. The constraint relates (via a STATE RESOURCE) STATE TASKs that cause the change in the property's state and PRECONDITION TASKs that specify the tests. For your problem, we need two STATE RESOURCE CONSTRAINTs, PAIN and PERFORMERS, that support discrete numerical domains and integrate STATE TASKs by adding their pain/performers values. For example, the PAIN STATE RESOURCE CONSTRAINT's projection of your pain will be increased by 4 if a STATE TASK that has a *Contribution* variable with a value of 4 is connected.

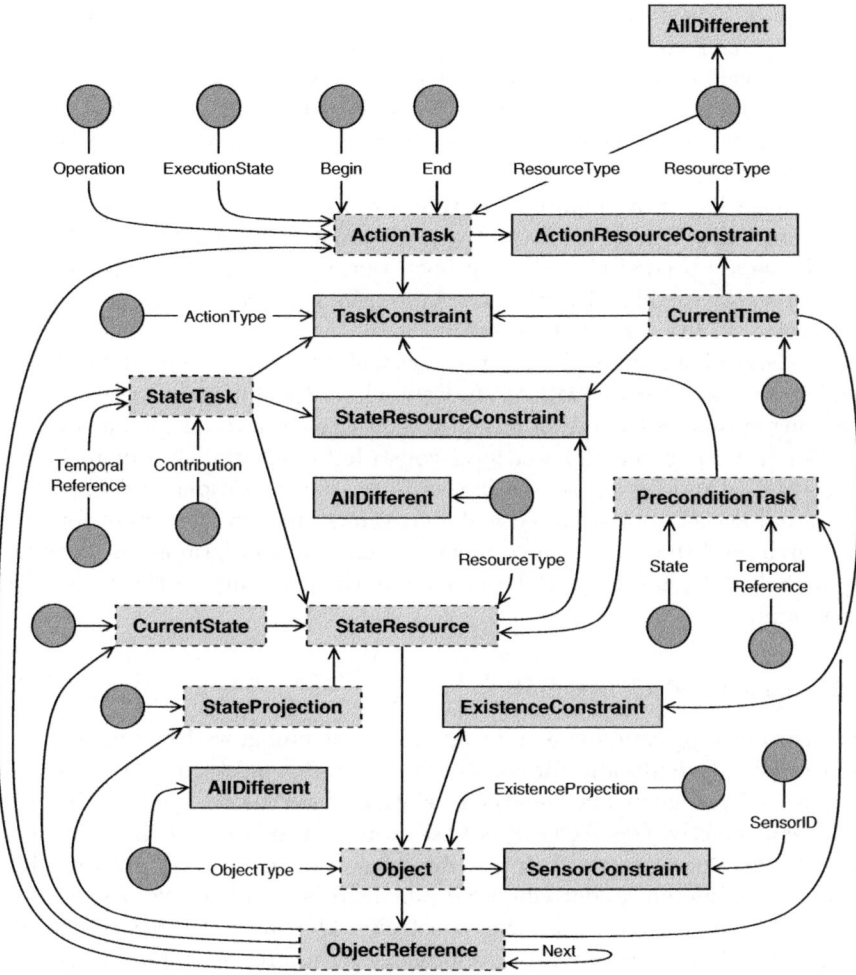

Fig. 4. Possible Type Relations

A TASK CONSTRAINT represents an action by specifying a relation between PRECONDITION TASKS, ACTION TASKS and/or STATE TASKS. It ensures that the connected tasks represent a valid action by enforcing a certain task configuration (i.e., the constraint specifies which tasks are to be related to the TASK CONSTRAINT and what kind of restrictions apply to the tasks' variables). For your problem, four different configurations satisfy a TASK CONSTRAINT. These express the three actions catch_only_one, catch_a_group and deliver_humans, each of them having an ACTION TASK for the ORC ACTION RESOURCE CONSTRAINT, a STATE TASK for the PAIN STATE RESOURCE CONSTRAINT and a STATE TASK for the PERFORMERS STATE RESOURCE CONSTRAINT. A fourth configuration consists of a PRECONDITION TASK for the PERFORMERS STATE

RESOURCE CONSTRAINT to establish the satisfaction goal `Performers` $\geq$ 5. In addition, the TASK CONSTRAINT requires that its tasks' variables have the appropriate values (e.g., if the TASK CONSTRAINT's *ActionType* variable has a value of `catch_a_group`, a value of 4 is required for the *Contribution* variable of the STATE TASK that is to be connected to the PAIN STATE RESOURCE CONSTRAINT).

Because it is not required for your example, I will not detail the role of constraints like OBJECTs, EXISTENCE CONSTRAINTs, SENSOR CONSTRAINTs, etc. These are needed to realize an open-world planning in which the type and number of the world's objects is not known in advance and can change without any action on the part of the agent.

You may have noticed that the constraints are not typical low-level constraints such as LESS or EQUALS. Indeed, a key aspect of the EXCALIBUR agent's planning system is the use of so-called *global constraints*. A global constraint is a substitute for a set of lower-level constraints, additional domain knowledge allowing the application of specialized data representations and algorithms to guide and accelerate search. Global constraints are very important for speeding up search, and the success of commercial constraint programming tools like the ILOG Scheduler [Le Pape 1994] can be ascribed mainly to their use of global constraints.

4.2   Structural Constraints

If constraint programming is to be applied to planning, we face the problem that conventional formulations for constraint satisfaction problems are too restrictive because all the elements and their relations have to be specified in advance, e.g., that exactly *five* ACTION TASKs would be related to the ORC ACTION RESOURCE CONSTRAINT. But possible plans can be structurally very different and require different underlying CSPs, e.g., CSPs with five, six or seven ACTION TASKs that are related to the ORC constraint. Of course, it is possible to create a maximal CSP that incorporates all possible structures as subproblems, which is similar to what most resource-based planning systems do with their plan-length bounds. But the problem here is that maximal structures scale *very* badly and are only feasible for small and only slightly variable structures, which is not the case with planning problems. And as we don't want to be restricted to plans of a certain length, this is not an option. Thus, an extension of CSP formulations must be used that includes the search problem for the CSP structure and is not tied to a special procedure like an extension along an increasing plan length.

A so-called *structural constraint satisfaction problem (SCSP)* can be used to overcome these deficiencies. In an SCSP, the constraint graph is not explicitly given. Only the types of constraints (together with their possible relations) and a set of *structural constraints* that restrict certain graph constellations are specified. A solution to the constraint problem is a constraint graph with an arbitrary number of variables, conventional (or object) constraints and connecting arrows, such that the graph and the variable assignments not only satisfy the conventional constraints (such as an ACTION RESOURCE CONSTRAINT) but also the structural constraints. The formalism is based on algebraic graph grammars [Rozenberg 1997].

Figure 5 shows an example of a structural constraint that checks if every ACTION TASK is linked to an ACTION RESOURCE CONSTRAINT with the same *ResourceType* variable. This prevents invalid ACTION TASK assignments, e.g., the ACTION TASK of walking being scheduled by the ACTION RESOURCE CONSTRAINT of the mouth. If the left-hand side of a structural constraint matches the constraint graph at some point, one of its right-hand sides (only one in the figure) has to match, too. For more details on structural constraints and SCSPs, please refer to [Nareyek 1999a] and [Nareyek 1999b]. A complete description of the structural constraints required for planning can be found in [Nareyek 2000].

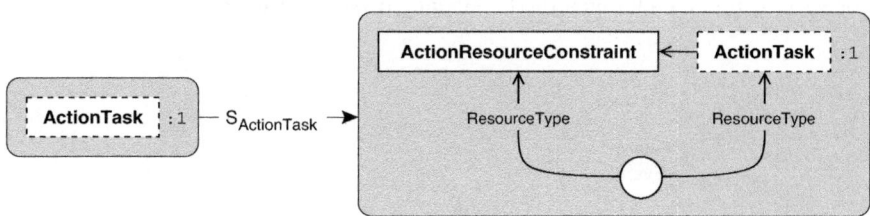

Fig. 5. The Structural Constraint ACTION TASK

The SCSP formulation is used to automatically derive rules/productions that can be applied to change the constraint graph during search. Thus, the search for valid variable assignments and for a valid graph structure can be seamlessly integrated.

But we have only talked about the specification of planning problems. Now I'm going to tell you how local search is used to find a solution to the problem."

4.3   Local Search

"Local search?" you say rather disdainfully because you have read somewhere that local search is something quite unscientific. Luckily, the annoyed wizard realizes that you are at present the only person he has to talk to (the other wizard appears to have left, presumably to get more of the great ethanol potion) and abstains from casting a teleportation spell on you to remedy the situation. "As I already mentioned," he resumes, "we have an undecidable/infinite search space where we can neither find all solutions nor verify that no solution exists. So the advantages of complete refinement search can't be exploited. On the other hand, the fact that local search is an anytime algorithm and that the techniques are not normally affected by dynamic changes in the search space make local search the ideal candidate for the computer-games domain. We live in a highly dynamic world and must react in real time!"

To cover up your previous faux pas, you grunt in assent. The wizard gives you a rather disconcerted look, but quickly collects himself and continues: "Another advantage of local search is that it is inherently a partial constraint satisfaction [Freuder & Wallace 1992]. This means that it is possible to find reasonable plans even if the problem is over-constrained, i.e., if it is not possible to create a fully correct plan. An extension of refinement search to handle partial constraint satisfaction would be much more inefficient, as the search space grows vastly.

Let's move on to how local search is accomplished. Local search means that there is always a complete assignment for the variables, and this assignment is iteratively changed toward a satisfactory/optimal solution. The quality of a current solution is measured by a so-called *objective function*. In contrast to an off-line refinement search, the iterative repair/improvement steps of local search make it possible to easily interleave sensing, planning and execution. After each single repair step, which can usually be computed very quickly, the planning process can involve changes in the situation (see Figure 6).

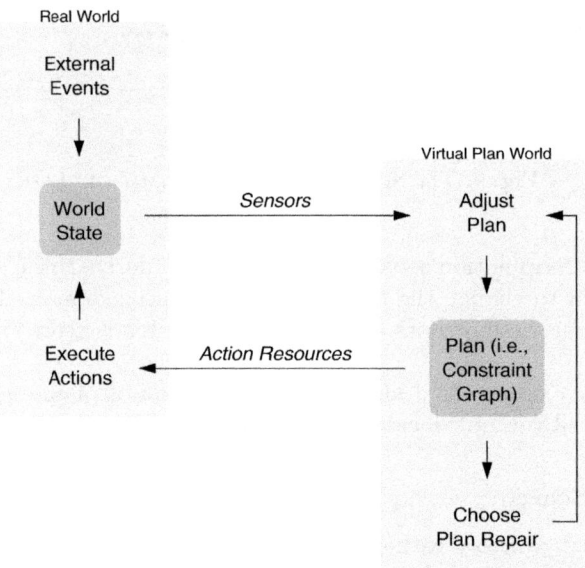

Fig. 6. Interleaving Sensing, Planning and Execution

The approach of [Nareyek 2001] is applied in the EXCALIBUR agent's planning system, which focuses the constraints to cause plan improvements. Each type of constraint can have a constraint-specific cost and goal function. Using the *cost function*, a constraint computes its inconsistency value, which expresses the degree of misassignment of the related variables. In connection with structural constraint satisfaction, invalid graph structures can also be penalized by the cost function (see [Nareyek 1999b]). Using the *goal function*, a constraint computes a preference value, which is used for optimization purposes.

For your problem, the PERFORMERS constraint's cost function is initialized to return the current plan's number of missing performers (to satisfy the PRE-CONDITION TASK of a TASK CONSTRAINT with an *ActionType* variable that represents the TASK CONSTRAINT's fourth configuration), the PAIN constraint's goal function to return the plan's resulting pain, and the ORC constraint's goal function to return the plan's total duration.

In addition, a constraint has *constraint-specific heuristics* that can change the values of the variables involved in order to reduce the constraint's inconsistency or improve its goal function value, e.g., changing a *Begin* variable to move a task to a better position in time. Domain-specific knowledge can be incorporated into the heuristics. In connection with structural constraint satisfaction, constraints can also use productions to change the graph in order to make an improvement, e.g., to add resources or actions. Thus, the search proceeds through a space consisting of complete but not necessarily optimal or correct plans.

4.4   The Constraints' Heuristics

Your problem does not require an application of sophisticated heuristics. The improvement heuristics can be the same for the PERFORMERS, PAIN and ORC constraint. To change the plan (i.e., the CSP), there are six modification alternatives[3]:

- add a `catch_only_one` TASK CONSTRAINT
- delete a `catch_only_one` TASK CONSTRAINT
- add a `catch_a_group` TASK CONSTRAINT
- delete a `catch_a_group` TASK CONSTRAINT
- add a `deliver_humans` TASK CONSTRAINT
- delete a `deliver_humans` TASK CONSTRAINT

For all six decision alternatives, there is a constraint-internal value that represents the preference for this alternative. All preference values are initially set to 1. If a constraint is called to effect an improvement of the plan, it increases an alternative's value by one if this alternative was chosen by the constraint's last improvement decision and the constraint's goal-/cost-function value is now better than the last time the constraint was called. If the goal-/cost-function value has deteriorated or is the same as last time, the preference value is decreased by two. Each time there is a consecutive deterioration, the decrease is doubled. However, no preference value can fall below one. Then, an alternative is chosen with a probability that is proportional to the alternative's preference value.

Let me illustrate the heuristic using the PAIN STATE RESOURCE CONSTRAINT. Consider a situation in which the current plan consists of 5 `catch_only_one` actions (C1A), 10 `catch_a_group` actions (CGA) and 3 `deliver_humans` actions (DHA) and the PAIN constraint is chosen to perform a plan improvement (see Situation 1 in Figure 7).

---

[3] A TASK CONSTRAINT is always added/deleted together with its required tasks.

Current Plan			Pain State Resource Constraint							
			Goal Function Value	Preference Values						Penalty
#C1A	#CGA	#DHA		+C1A	-C1A	+CGA	-CGA	+DHA	-DHA	
Situation 1										
5	10	3	12	1	2	1	3	3	1	2
				1	2	1	3	4	1	2
						select +DHA				
5	10	4								
...										
Situation 2										
5	14	4	17	1	2	1	3	4	1	2
				1	2	1	3	2	1	4
						select +DHA				
5	14	5								
...										
Situation 3										
8	16	5	17	1	2	1	3	2	1	4
				1	2	1	3	1	1	8
						select -CGA				
8	15	5								
...										
Situation 4										
11	15	5	16	1	2	1	3	1	1	8
				1	2	1	4	1	1	2

Fig. 7. The Heuristic of the PAIN STATE RESOURCE CONSTRAINT

The PAIN constraint's goal function calculates the sum of the *Contribution* variables of the linked STATE TASKs. There are currently 5 linked STATE TASKs with a *Contribution* of 1, 10 with a *Contribution* of 4, and 3 with a *Contribution* of -11. Thus, the constraint's goal function has a value of $5 \times 1 + 10 \times 4 + 3 \times (-11) = 12$. The current preference values of the constraint are given in Figure 7 ("+" means addition and "-" means deletion).

Let's assume that the constraint's goal function value is now better than the last time the constraint was called and that the constraint's last improvement decision was to add a `deliver_humans` action. Accordingly, this decision's preference value is rewarded by increasing it by one (next line in Figure 7). Next, an alternative to improve the current situation is selected. The choice probability for the +DHA option is the highest (preference value divided by the sum of all preference values; $4/12 = 33\%$), and we assume that this alternative is chosen. The plan/CSP is changed according to this option.

After some iterations, the PAIN constraint can be called again (see Situation 2 in Figure 7). In the meantime, the plan has been changed and some more catch_a_group actions are added. The PAIN constraint's goal function value has deteriorated since the last time the constraint was called and its last decision's preference value is therefore decreased by the penalty value of 2. In addition, the penalty value is doubled. The choice probability for the +DHA option is now only $2/10 = 20\%$, but we assume that this alternative is chosen again. The plan is changed according to this option.

After some time, the PAIN constraint is called once more (Situation 3). The constraint's goal function value is the same as at the last call. Stagnation is considered to be as bad as deterioration, and the +DHA option's preference value is therefore decreased by the penalty value of 4, and the penalty value is doubled again. However, the +DHA preference value is increased to its minimum of 1 to ensure that the option retains a chance of being chosen. We assume that the -CGA alternative is chosen this time (probability of $3/9 = 33\%$). The plan is changed according to this option.

At the next call of the PAIN constraint (Situation 4), the constraint's goal function value has improved, and so, the -CGA preference value is increased and the penalty value is set back to 2.

4.5   The Global Search Control

As described above, global constraints have integrated heuristics to enable them to choose a successor state on their own. For each iteration of local search, a so-called *global search control* selects the constraint that is to perform the change. The choice depends on the constraints' cost and goal function values and implements the goal criteria.

The cost and goal functions may compete such that an improvement of one function value may lead to a deterioration in the other function value. Thus, various options for the selection of a function for improvement are customizable, e.g., alternating phases of a predefined length or with special abort criteria, or a general cost measure consisting of the addition of the single function values multiplied by static (or even variable) coefficients. A constraint that is selected for improvement is told whether to improve its goal or its cost function.

For your problem, within the global search control, we will always choose the PERFORMERS constraint to perform an improvement if the plan does not yield at least 5 performers. Otherwise, if the plan yields any pain, we'll choose the PAIN constraint. If we have a plan with enough performers and no pain, we'll choose the ORC constraint to help shorten the plan. Thereby, we can realize a search that has the performers criterion as satisfaction goal, the pain criterion as the primary optimization goal and the duration criterion as the secondary optimization goal.

The global search control also serves as a general manager, which variables and constraints can dock on to and off from in a plug-and-play manner. This is necessary if the constraint graph is changed by heuristics that make structural changes and it provides a simple mechanism for tackling problems that involve dynamic changes.

We'll initialize the search process with a CSP that includes the PERFORMERS, PAIN and ORC constraint and a TASK CONSTRAINT with an *ActionType* variable that represents the TASK CONSTRAINT's fourth configuration (to establish the satisfaction goal `Performers` $\geq$ 5). The STATE RESOURCE CONSTRAINTS PERFORMERS and PAIN are initialized with *CurrentState* values of 0.

During search, the structural constraints have to be enforced in the same way as the other constraints. But if one can prove that the potential changes in the constraints' heuristics cannot violate any structural constraints, this can be skipped. For example, the additions/deletions of actions in our constraints' decision alternatives can easily be performed in a way that do not threaten structural constraints.

4.6   Solving the Orc Quest Problem

If you now start the planning system, you will quickly obtain the shortest possible plan that results in no pain for you (55 times `catch_only_one` and 5 times `deliver_humans`, which yields a plan duration of **115 hours**, a **0 on the pain scale** and **5 performers**). Figure 8 shows the temporal distribution for 100,000 test runs with different random seeds (e.g., after 1,000 improvement iterations, 100 % of the test runs found a plan that yields enough performers, 94 % of the test runs found a plan that yields enough performers and no pain, and 41 % of the test runs found the shortest plan that yields enough performers and no pain)."

"That looks nice," you comment, "but what about a comparison with other planning systems?" The wizard shakes his head: "Don't you remember the problems you had? The other systems will either come up with the plan that primarily minimizes the plan length (`catch_a_group` & `catch_a_group`) or require a bound for the plan length. If we cheat in favor of the other planning systems and provide the information that the optimal length bound is 60 actions, an encoding according to CPlan [Van Beek & Chen 1999] of this simple problem would require 129,600 variables[4] — not to speak of constraints! Even the time required to create a list of these domain variables (tested with CHIP, which is a state-of-the-art constraint programming system) is longer than any of the test runs in Figure 8 needed to compute the optimal plan. However, we actually cheated in favor of CPlan by providing the optimal length bound. In general, the comparison is not a quantitative question of speed but a qualitative question of *capability* to find the optimum.

---

[4] **60** steps × (duration domain **60** × **6** + pain domain **60** × **16** + performers domain **60** × **14**) = **129,600**.

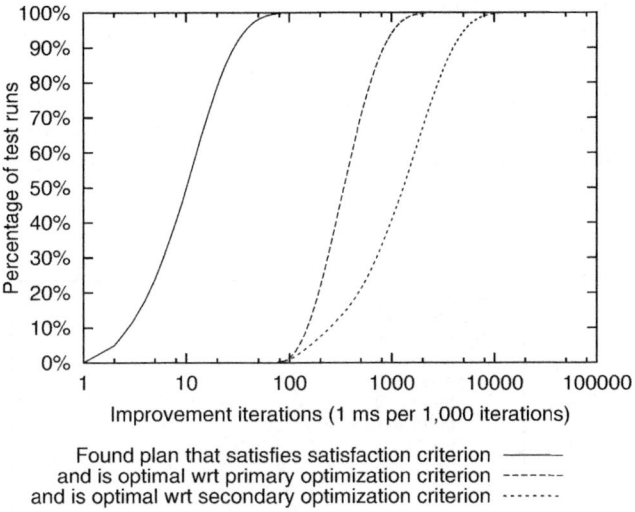

Found plan that satisfies satisfaction criterion ———
and is optimal wrt primary optimization criterion -------
and is optimal wrt secondary optimization criterion ········

Fig. 8. Test-Run Results for the Orc Quest Example

However, if you insist on quantitative results, we can start to improve the constraints' heuristics. As I said, domain-dependent knowledge can easily be integrated. The heuristics are quite general so far and don't make much use of your specific problem's characteristics. For example, we could improve the PAIN constraint by paying more attention keeping the same number of performers if an option to reduce the pain is chosen. For this purpose, three additional catch_on-ly_one actions can be added if the option to delete a catch_a_group action is chosen. In the same way, the number of performers can be replenished by adding catch_only_one and catch_a_group actions (partitioning them according to the preference values) if the option of adding a deliver_humans action is chosen. Even such simple modifications make it possible to reduce execution times by 20 % (compared to the results in Figure 8). However, the selection mechanisms involved can also be improved. For example, if the constraint-internal choice of a modification option is changed, such that the option with the highest preference value is chosen (instead of making a choice based on a probabilistic distribution), a further reduction of 20 % can be achieved.

5   Related Work and Conclusion

I have explained the approach of the EXCALIBUR agent's planning system. Your problem was a very simple one but it allowed me to outline the system's basic features and show why existing approaches are not able to handle it adequately. Specialized ACTION and STATE RESOURCE CONSTRAINTs were applied to your problem — of course, for more sophisticated problems involving precondition requirements and temporal aspects, more general versions must be available. More general / domain-independent versions of the planning model's constraints have already been developed and successfully tested on problems like the Logistics Domain. The constraints' heuristics also include operations that take into account the temporal distribution, like temporally shifting tasks to more suitable time points. Figure 9 shows run-time distributions for some simple problems of the AIPS-2000 planning competition (Track 2). I will report on this in more detail soon.

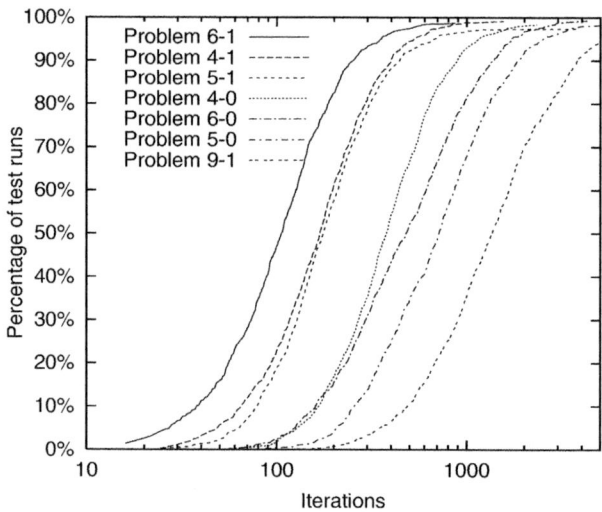

Fig. 9. Run-Time Distributions for the Logistics Domain

For planning problems that involve resource constraints, some planning systems keep planning and resource allocation separate, e.g., the approach of [Srivastava & Kambhampati 1999]   and   the *parc*PLAN   system   by [El-Kholy & Richards 1996]. The separation of planning and resource allocation prevents the systems from considering interactions between the decisions with respect to planning and resource assignment, which is a great disadvantage if resource-related properties are to be optimized. Consequently, the resources serve only as constraints for the planning problem and are not used as primary optimization goals. The same applies to resource-based planning systems that integrate planning and resource allocation, like O-Plan2 [Drabble & Tate 1994], IxTeT [Laborie & Ghallab 1995], the LPSAT engine's application to planning [Wolfman & Weld 1999] and IPP's extension [Koehler 1998]. All of these focus primarily on optimization of the plan length, which is a rather curious approach as this property is usually irrelevant.

Resource-based planning systems that are based entirely on general search frameworks like CP or OR require bounds for the plan length or the number of actions. If a correct plan cannot be found, these bounds can be expanded. Some examples here are the OR-based approach of [Bockmayr & Dimopoulos 1998], ILP-PLAN [Kautz & Walser 1999], CPlan [Van Beek & Chen 1999] and the approach of [Rintanen & Jungholt 1999]. Again, these systems primarily optimize the plan-length property. An optimum with respect to resource-related optimization goals can only be found if the initial bound can be set such that the optimal solution is guaranteed to lie within this bound. This is a very hard task for specific problems and impossible at a general level. Besides, creating the maximal structures for the search space is much too costly for complex real-world problems."

"What do you mean by *real world*?" you put in doubtfully. The wizard quickly corrects himself: "Sorry, I meant of course: for complex virtual-world problems.

Most planning systems use highly specific representations and algorithms, which makes it very hard to integrate new features like reasoning about resources. Unlike these planning formalisms, the EXCALIBUR agent's planning system is based on the general search framework of constraint programming. In contrast to other general search frameworks such as propositional satisfiability (SAT) and OR (i.e., integer linear programming), the CP framework allows the use of higher-level constraints. The propositional clauses of SAT and the linear inequations of OR can't exploit higher-level domain knowledge to support search. Besides, the SAT approach doesn't have numbers in its representation repertoire. Another reason for not using OR is that there are lots of planning decisions with discrete alternatives, and the performance of OR methods declines sharply as the number of integer variables increases.

The concept of using global constraints for local search was applied, which enables us to integrate domain-dependent knowledge using the global constraints in a plug-and-play manner. As the search space is undecidable, this heuristic knowledge is essential to guide and accelerate search. Other local search approaches to planning, like SATPLAN [Kautz & Selman 1996], GPG [Gerevini & Serina 1999] and LPSP [Brafman & Hoos 1999], do not use domain-dependent knowledge. The closest relative to the EXCALIBUR agent's planning is the ASPEN system [Chien et al. 2000,Rabideau et al. 1999][5]. This system uses local search for resource-restricted planning problems in a similar way. However, the sensing and handling of an open world is somewhat limited and the system is not based on a general search framework like constraint programming.

To make the search independent of the focus on plan length, the paradigm of structural constraint satisfaction was used. This extension allows the search to make all possible kinds of changes in the constraint graph during the search process — not only after the current graph has been completely searched, and not only along an increasing plan length. Unlike other planning systems that use productions/rules to change the plan as part of the search (e.g., [Ambite & Knoblock 1997]), modeling planning as an SCSP allows us to specify the problem in a declarative manner and enables the corresponding productions to be deduced automatically. The automatic method guarantees that local-search

---

[5] See also other articles in this book.

methods can potentially find all valid plans, which is not normally the case with manual solutions.

The planning model of the EXCALIBUR agent's planning system borrows from typical constraint-based applications for resource allocation/optimization. The power of global constraints for constraint-specific representation and reasoning mechanisms for specific resource types was recognized here very early on and led to significant speedups in the solution process. General frameworks for planning and scheduling like HSTS [Muscettola 1994] lack such specialized representation and reasoning capabilities.

In conclusion, we can say that the EXCALIBUR agent's planning system conducts the search in a way that focuses on the optimization of resource-related properties. Even though the model supports full domain-independent planning, domain-specific heuristic knowledge can easily be integrated. The local search approach enables the system to provide very primitive plans (reactions) for short-term computation horizons, while longer computation times are used to improve and optimize the agent's behavior. The system's properties make it perfectly suitable for the application domain of computer games. Not only genres like computer role-playing games and strategy games can benefit from this. For games like first-person shooters, too, increasing efforts are being made to include a sophisticated behavior for their bots."

The wizard appears to have no more to say, so you chip in with a comment designed to keep him talking until you have drained your jar: "I've heard of something called hierarchical planning." Obviously, you've struck it lucky because the wizard answers gleefully: "Now that's interesting! The hierarchical planning approach has some features in common with the EXCALIBUR agent's planning system. Hierarchical planning is not bound to the plan length either and is also a kind of heuristic search. But the refinement hierarchy is usually created manually and rarely has the potential to construct every possible solution. Actually, I'm confident that the EXCALIBUR agent's planning system will soon be extended by hierarchical planning mechanisms. The hierarchy would set a general heuristic direction, allowing the iterative repair to vary the plan as needed."

Just as you are about to leave the wizard to set off on your hunt for humans, you have the sudden inspiration that more information on the underlying EXCALIBUR project is available at:

`http://www.ai-center.com/projects/excalibur/`

You have no idea what brought this on or what it means and put it down to an undocumented subfunction of your behavior program, triggered by the approach of the acknowledgments.

Acknowledgments. The work reported here is supported by the German Research Foundation (DFG), NICOSIO, Conitec Datensysteme GmbH and Cross Platform Research Germany (CPR).

References

[Ambite & Knoblock 1997] Ambite, J. L., and Knoblock, C. A. 1997. Planning by Rewriting: Efficiently Generating High-Quality Plans. AAAI-97, 706–713.

[Blum & Furst 1997] Blum, A. L., and Furst, M. L. 1997. Fast Planning Through Planning Graph Analysis. *Artificial Intelligence* 90: 281–300.

[Bockmayr & Dimopoulos 1998] Bockmayr, A., and Dimopoulos, Y. 1998. Mixed Integer Programming Models for Planning Problems. In Working Notes of the CP98 Workshop on Constraint Problem Reformulation.

[Boutilier, Dean & Hanks 1999] Boutilier, C.; Dean, T.; and Hanks, S. 1999. Decision-Theoretic Planning: Structural Assumptions and Computational Leverage. *Journal of Artificial Intelligence Research* 11: 1–94.

[Brafman & Hoos 1999] Brafman, R. I., and Hoos, H. H. 1999. To Encode or not to Encode – I: Linear Planning. IJCAI-99, 988–993.

[Chien et al. 2000] Chien, S.; Rabideau, G.; Knight, R.; Sherwood, R.; Engelhardt, B.; Mutz, D.; Estlin, T.; Smith, B.; Fisher, F.; Barrett, T.; Stebbins, G.; and Tran, D. 2000. ASPEN — Automating Space Mission Operations using Automated Planning and Scheduling. In Proceedings of the Sixth International Symposium on Technical Interchange for Space Mission Operations and Ground Data Systems (SpaceOps 2000).

[Drabble & Tate 1994] Drabble, B. and Tate, A. 1994. The Use of Optimistic and Pessimistic Resource Profiles to Inform Search in an Activity Based Planner. AIPS-94, 243–248.

[El-Kholy & Richards 1996] El-Kholy, A., and Richards, B. 1996. Temporal and Resource Reasoning in Planning: the *parc*PLAN approach. ECAI-96, 614–618.

[Ephrati, Pollack & Milshtein 1996] Ephrati, E.; Pollack, M. E.; and Milshtein, M. 1996. A Cost-Directed Planner: Preliminary Report. In Proceedings of the Thirteenth National Conference on Artificial Intelligence (AAAI-96), 1223–1228.

[Erol, Nau & Subrahmanian 1991] Erol, K.; Nau, D. S.; and Subrahmanian, V. S. 1991. Complexity, Decidability and Undecidability Results for Domain-Independent Planning. Technical Report CS-TR-2797, University of Maryland, Institute for Advanced Computer Studies, Maryland, USA.

[Freuder & Wallace 1992] Freuder, E. C., and Wallace, R. J. 1992. Partial Constraint Satisfaction. *Artificial Intelligence* 58: 21–70.

[Gerevini & Serina 1999] Gerevini, A., and Serina, I. 1999. Fast Planning through Greedy Action Graphs. AAAI-99, 503–510.

[ILOG, Inc. 2000] ILOG, Inc. 2000. Optimization Technology White Paper — A comparative study of optimization technologies. White Paper, ILOG, Inc., Mountain View, CA.

[Kautz & Selman 1996] Kautz, H., and Selman, B. 1996. Pushing the Envelope: Planning, Propositional Logic, and Stochastic Search. AAAI-96, 1194–1201.

[Kautz & Walser 1999] Kautz, H., and Walser, J. P. 1999. State-space Planning by Integer Optimization. AAAI-99, 526–533.

[Koehler 1998] Koehler, J. 1998. Planning under Resource Constraints. ECAI-98, 489–493.

[Laborie & Ghallab 1995] Laborie, P., and Ghallab, M. 1995. Planning with Sharable Resource Constraints. IJCAI-95, 1643–1649.

[Le Pape 1994] Le Pape, C. 1994. Implementation of Resource Constraints in ILOG Schedule: A Library for the Development of Constraint-Based Scheduling Systems. *Intelligent Systems Engineering* 3(2): 55–66.

[Mattsson 2000] Mattsson, C. 2000. The Tolkien Monster Encyclopedia.
http://home7.swipnet.se/~w-70531/Tolkien/

[Muscettola 1994] Muscettola, N. 1994. HSTS: Integrating Planning and Scheduling.
In Zweben, M., and Fox, M. S. (eds.), *Intelligent Scheduling*, Morgan Kaufmann,
169–212.

[Nareyek 1999a] Nareyek, A. 1999. Structural Constraint Satisfaction. In Papers from
the AAAI-99 Workshop on Configuration, 76–82. (available via the EXCALIBUR
project's webpage)

[Nareyek 1999b] Nareyek, A. 1999. Applying Local Search to Structural Constraint
Satisfaction. In Proceedings of the IJCAI-99 Workshop on Intelligent Workflow and
Process Management: The New Frontier for AI in Business. (available via the EX-
CALIBUR project's webpage)

[Nareyek 2000] Nareyek, A. 2000. Open World Planning as SCSP. In Papers from the
AAAI-2000 Workshop on Constraints and AI Planning, Technical Report, WS-00-02,
35–46. AAAI Press, Menlo Park, California. (available via the EXCALIBUR project's
webpage)

[Nareyek 2001] Nareyek, A. 2001. Using Global Constraints for Local Search. In
Freuder, E. C., and Wallace, R. J. (eds.), *Constraint Programming and Large Scale
Discrete Optimization*, American Mathematical Society Publications, DIMACS Vol-
ume 57, 9–28. (available via the EXCALIBUR project's webpage)

[Rabideau et al. 1999] Rabideau, G.; Knight, R.; Chien, S.; Fukunaga, A.; and Govindjee, A. 1999. Iterative Repair Planning for Spacecraft Operations in the ASPEN
System. International Symposium on Artificial Intelligence Robotics and Automation in Space (ISAIRAS).

[Rintanen & Jungholt 1999] Rintanen, J., and Jungholt, H. 1999. Numeric State Variables in Constraint-based Planning. ECP-99.

[Rozenberg 1997] Rozenberg, G. ed. 1997. *The Handbook of Graph Grammars. Volume
I: Foundations.* Reading, World Scientific.

[Srivastava & Kambhampati 1999] Srivastava, B., and Kambhampati, S. 1999. Scaling
up Planning by teasing out Resource Scheduling. ECP-99.

[Van Beek & Chen 1999] Van Beek, P., and Chen, X. 1999. CPlan: A Constraint Programming Approach to Planning. AAAI-99, 585–590.

[Williamson & Hanks 1994] Williamson, M., and Hanks, S. 1994. Optimal Planning
with a Goal-Directed Utility Model. In Proceedings of the Second International
Conference on AI Planning Systems (AIPS-94), 176–181.

[Wolfman & Weld 1999] Wolfman, S. A., and Weld, D. S. 1999. The LPSAT Engine &
its Application to Resource Planning. In Proceedings of the Sixteenth International
Joint Conference on Artificial Intelligence (IJCAI-99), 310–316.

An Empirical Evaluation of the Effectiveness of Local Search for Replanning⋆

Steve Chien, Russell Knight, and Gregg Rabideau

Jet Propulsion Laboratory
California Institute of Technology
4800 Oak Grove Drive
Pasadena, CA 91109
{chien, knight, rabideau}@aig.jpl.nasa.gov

Abstract. Local search has been proposed as a means of responding to changes in problem context requiring replanning. Iterative repair and iterative improvement have desirable properties of preference for plan stability (e.g., non-disruption, minimizing change), and have performed well in a number of practical applications. However, there has been little real empirical evidence to support this case. This paper focuses on the use of local search to support a continuous planning process (e.g., continuously replanning to account for problem changes) as is appropriate for autonomous spacecraft control. We describe results from ongoing empirical tests using the CASPER system to evaluate the effectiveness of local search to replanning using a number of spacecraft scenario simulations including landed operations on a comet and rover operations.

1 Introduction

In recent years Galileo, Clementine, Mars Pathfinder, Lunar Prospector, and Cassini have all demonstrated a new range of robotic missions to explore our solar system. However, complex missions still require large teams of highly knowledgeable personnel working around the clock to generate and validate spacecraft command sequences. Increasing knowledge of our Earth, our planetary system, and our universe challenges NASA to fly large numbers of ambitious missions, while fiscal realities require doing so with budgets far smaller than in the past. In this climate, the automation of spacecraft commanding becomes an endeavor of crucial importance.

Automated planning is a key enabling technology for autonomous spacecraft. Recent experiences indicate the promise of planning and scheduling technology for space operations. Use of the DATA-CHASER automated planning and scheduling system (DCAPS) to command the DATA-CHASER shuttle payload reduced commanding-related mission operations effort by 80% and increased science return by 40% over manually generated sequences [1]. This increase was

---

⋆ This work was performed by the Jet Propulsion Laboratory, California Institute of Technology, under contract with the National Aeronautics and Space Administration.

A. Nareyek (Ed.): Local Search for Planning and Scheduling, LNAI 2148, pp. 79–94, 2001.
© Springer-Verlag Berlin Heidelberg 2001

possible because short turn-around times (approximately 6 hours) imposed by operations constraints did not allow for lengthy, manual optimization. And the Remote Agent Experiment [2] demonstrated the feasibility of flying AI software (including a planner) to control a spacecraft.

Local iterative algorithms have been successfully applied to planning and scheduling [1,3,4,5] for a wide range of space applications. Local iterative repair has been proposed as a means of providing a fast replanning capability to enable response to environmental changes [6,7]. This paper describes an empirical evaluation of the effectiveness of local search in such a replanning context.

The remainder of this paper is organized as follows. First, we briefly describe our approach to local iterative repair. Next we describe how it is applied in a replanning context. We then describe a Comet Nucleus Sample Return (CNSR) scenario and simulation and empirical tests evaluating the effectiveness of local search in finding solutions. Finally, we describe future work, related work and conclusions.

2   Plan Conflicts and Repair

We now describe the overall approach to iterative repair planning and scheduling in the ASPEN system [5,8]. The ASPEN planning and scheduling system is able to represent a wide range of constraints including:

- Finite enumeration state requirements and changers (e.g., camera ON, OFF);
- Depletable (e.g., fuel) and non-depletable (e.g., 20W power) resource constraints;
- Task decompositions (e.g., Hierarchical Task Networks);
- Complex functional relationships between activity parameters (e.g., downlink time is data/rate + startup); and
- Metric time constraints (e.g., calibrate the camera 20-30s before the observation).

ASPEN supports a suite of search engines to support integrated planning and scheduling to produce activity plans that satisfy the constraints. The remainder of this section briefly outlines the most commonly used search strategy - local iterative repair.

We define a conflict as a particular class of ways to violate a plan constraint (e.g., over-use of a resource or an illegal state transition). For each conflict type, there is a set of repair methods. The search space consists of all possible repair methods applied to all possible conflicts in all possible orders. We describe an efficient approach to searching this space. During iterative repair, the conflicts in the schedule are detected and addressed one at a time until no conflicts exist, or a user-defined time limit has been exceeded. A conflict is a violation of a parameter dependency, temporal or resource constraint. Conflicts can be repaired by means of several predefined methods. The repair methods are: moving an activity, adding a new instance of an activity, deleting an activity, detailing an activity, abstracting an activity, making a reservation of an activity, canceling a reservation, connecting a temporal constraint, disconnecting a constraint, and

changing a parameter value. The repair algorithm first selects a conflict to repair then selects repair method. The type of conflict being resolved determines
which methods can repair the conflict. Depending on the selected method, the
algorithm may need to make additional decisions. For example, when moving an
activity, the algorithm must select a new start time for the activity.

Note that resolving a conflict may cause another conflict so that complete
resolution of a set of similar conflicts may require several steps. For example, disconnecting a violated temporal constraint would remove the temporal constraint
violation but would introduce a new unconnected temporal constraint conflict.
This would then be resolved by creating a new activity and linking in or by
linking to an existing activity that correctly satisfies the temporal constraint.
Likewise many conflicts can be resolved by abstracting the offending activity.
However this causes an undetailed activity conflict and requires re-expansion of
the abstract activity. The abstract activity can then be detailed again in a way
that perhaps avoids the original conflict(s).

Figure 1 shows an example situation for repair. On-board RAM is represented as a depletable resource. The shaded region shows a conflict where the
RAM buffer has been oversubscribed. The science activities using the resource
prior to the conflict are considered contributors. Moving or deleting one of the
contributors can repair the conflict. Another possibility would be to create a new
downlink activity in order to replenish the resource and repair the conflict.

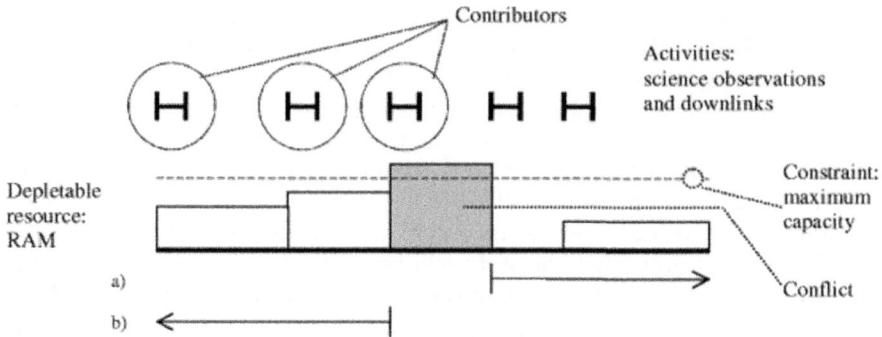

Fig. 1. Repairing a depletable resource conflict. The arrows show time intervals that
resolve the conflict by a) moving a positive contributor or b) adding a negative contributor.

3   Integrating Planning and Execution

Traditionally, planning and scheduling research has focused on a batch formulation of the problem. In this approach, when addressing an ongoing planning
problem, time is divided up into a number of planning horizons, each of which
lasts for a significant period of time. When one nears the end of the current

horizon, one projects what the state will be at the end of the execution of the current plan (see Figure 2). The planner is invoked with: a new set of goals for the new horizon, the expected initial state for the new horizon, and the planner generates a plan for the new horizon. As an exemplar of this approach, the Remote Agent Experiment operated in this fashion [9].

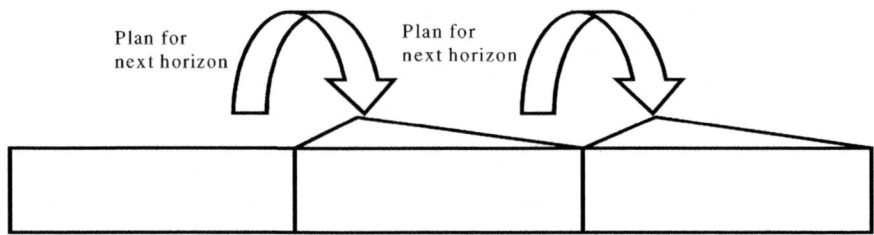

Fig. 2. Traditional Batch "Plan then Execute" Cycle

This approach has a number of drawbacks. In this batch oriented mode, typically planning is considered an off-line process which requires considerable computational effort and there is a significant delay from the time the planner is invoked to the time that the planner produces a new plan.[1] If a negative event occurs (e.g., a plan failure), the response time until a new plan may be significant. During this period the system being controlled must be operated appropriately without planner guidance. If a positive event occurs (e.g., a fortuitous opportunity, such as activities finishing early), again the response time may be significant. If the opportunity is short lived, the system must be able to take advantage of such opportunities without a new plan (because of the delay in generating a new plan). Finally, because the planning process may need to be initiated significantly before the end of the current planning horizon, it may be difficult to project what the state will be when the current plan execution is complete. If the projection is wrong the plan may have difficulty.

To achieve a higher level of responsiveness in a *dynamic planning* situation, we utilize a *continuous planning* approach and have implemented a system called CASPER (for Continuous Activity Scheduling Planning Execution and Replanning). Rather than considering planning a batch process in which a plan-

---

[1] As a data point, the planner for the Remote Agent Experiment (RAX) flying onboard the New Millennium Deep Space One mission [9] takes approximately 4 hours to produce a 3 day operations plan. RAX is running on a 25 MHz RAD 6000 flight processor and uses roughly 25% of the CPU processing power. While this is a significant improvement over waiting for ground intervention, making the planning process even more responsive (e.g., on a time scale of seconds or tens of seconds) to changes in the operations context, would increase the overall time for which the spacecraft has a consistent plan. As long as a consistent plan exists, the spacecraft can keep busy working on the requested goals and hence may be able to achieve more science goals.

ner is presented with goals and an initial state, the planner has a current goal set, a plan, a current state, and a model of the expected future state. At any time an incremental update to the goals, current state, or planning horizon (at much smaller time increments than batch planning)[2] may update the current state of the plan and thereby invoke the planner process. This update may be an unexpected event or simply time progressing forward. The planner is then responsible for maintaining a consistent, satisficing plan with the most current information. This current plan and projection is the planner's estimation as to what it expects to happen in the world if things go as expected. However, since things rarely go exactly as expected, the planner stands ready to continually modify the plan. From the point of view of the planner, in each cycle the following occurs:

- changes to the goals and the initial state first posted to the plan,
- effects of these changes are propagated through the current plan projections (includes conflict identification)
- plan repair algorithms[3] are invoked to remove conflicts and make the plan appropriate for the current state and goals.

This approach is shown in below in Figure 3.

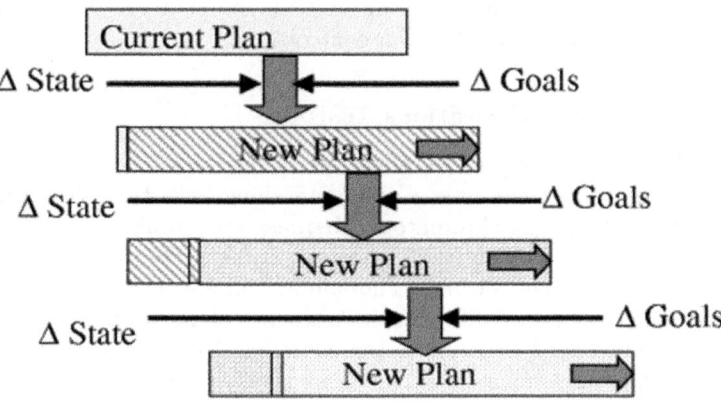

Fig. 3. Continuous Planning Incremental Plan Extension

At each step, the plan is created by using incremental replanning from:

- the portion of the old plan for the current planning horizon;
- the change (Δ) in the goals relevant for the new planning horizon;
- the change (Δ) in the state; and
- the new (extended) planning horizon.

---

[2] For the spacecraft control domain we are envisaging an update rate on the order of tens of seconds real time.

[3] In this paper we do not focus on the state/resource representation or the repair methods, for details see (Rabideau et al. 1999).

The key assumption of this approach is the belief that:

IF
     the change in the goals is small (Δ goals small)
     and the change in state is small (Δ state small)
THEN
     the change in the plan required to bring the plan consistent with the
     state and goals will be small (Δ plan small).

Of course, there is no guarantee that this hypothesis will be met. Indeed, if the goals in the domain are highly interacting (e.g., a small change in the combination of goals may require major changes to the plan). However, an important point to note is that the use of varying abstraction levels can aid greatly in making this incremental property hold. Often a plan modification that would require a large number of changes at a lower level of abstraction can be represented as a single change at the more abstract level (e.g., moving a group of activities associated with a single observation could be performed as a single move operation).

This incremental fast replanning approach as embodied in the CASPER system is being used in a range of applications [5] including: onboard a research prototype rover, planned for flight in several space missions, high level flight control and weapons management in an Unmanned Aerial Vehicle prototype, and Ground Communication Station control.

3.1   CNSR Landed Operations Testbed

The CNSR scenario represents landed operations of a mission to a comet (see Figure 4). The lander will use a one-meter long drill to collect samples and then feed them to a gas chromatograph/mass spectrometer onboard the lander. This instrument will analyze the composition of the nucleus collected from various depths below the surface. The lander will also carry cameras to photograph the comet surface. Additional instruments planned onboard the lander to determine the chemical makeup of the cometary ices and dust will include an infrared/spectrometer microscope and a gamma-ray spectrometer. After several days on the surface, the lander will bring a sample back to the orbiter for return to Earth.

In this test scenario the planner has models of 11 state and resource timelines, including drill location, battery power, data buffer, and camera state. The model also includes 19 activities such as uplink data, move drill, compress data, take picture, and perform oven experiment.

The nominal mission scenario consists of three major classes of activities: drilling and material transport, instrument activity including imaging and in-situ materials experiments, and data uplink. Of these, drilling is the most complex and unpredictable.

The mission plan calls for three separate drilling activities. Each drilling activity drills a separate hole and acquires samples at three different depths during the process: a surface sample, a 20 cm deep sample, and a one-meter deep sample. Acquiring a sample involves five separate "mining" operations after

Fig. 4. Artist depiction of ST4 lander landing on Comet

the hole has been drilled to the desired depth. Each mining operation removes 1 cm of material. Drilling rate and power are unknown a priori, but there are reasonable worst-case estimates available. Drilling can fail altogether for a variety of reasons.

One of the three drilling operations is used to acquire material for sample-return. The other two are used to supply material to in-situ science experiments onboard the lander. These experiments involve depositing the samples in an oven, and taking data while the sample is heated. Between baking operations the oven must cool, but there are two ovens, allowing experiments to be interleaved unless one of the ovens fails.

The replanning capability was tested using a stochastic version of the CNSR simulation described above. This simulation had a number of random variables, which are described below to show the richness of the domain.

- Compression - we model the compression for science data as a normal random variable with a mean of 0.9 and a standard deviation of 0.25*0.9. This has the effect of forcing the planner to respond to buffer over-runs (as described above) and buffer under-runs (to optimize the plan).
- Drilling Time - we model the amount of time to drill in minutes as a random variable with mean of 30 and standard deviation of 3.
- Drilling power - we model the actual power consumption from drilling in watts as a normal random variable with mean 40 and standard deviation 4.
- Oven Failure - we model oven failure occurrence as Poisson distributed with each oven having a 50% chance of failure over the entire mission horizon.
- Data Transmission Rate: we model the time to transmit data in kilobits per second as a normal random variable with a mean of 100 and a standard deviation of 10. This is intended to model the variability in communications to the orbiter.
- Oven Warming and Cooling Times: we model the amount of time to heat up the sample and for the oven to cool down in minutes as random variables with means of 30 and 120, and standard deviations of 3 and 12, respectively. This is intended to model the unknown thermal properties of the samples.

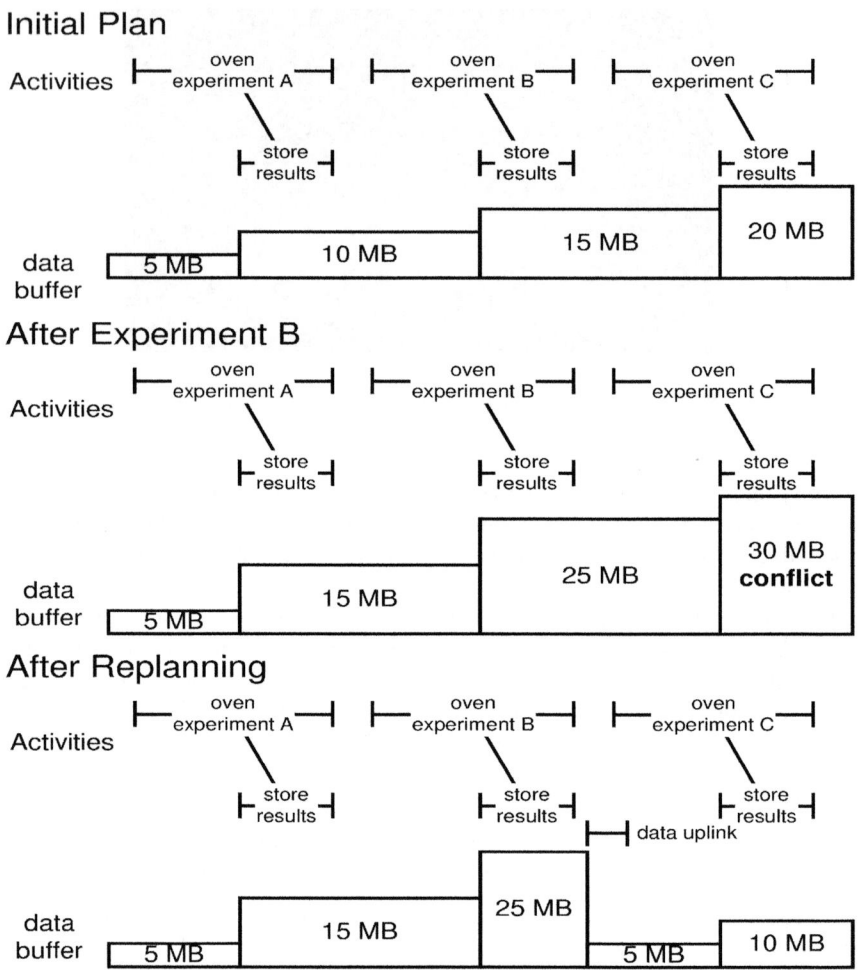

Fig. 5. Over-subscribed Data Buffer Example

To illustrate the operation of CASPER in the CNSR domain consider the following example of CASPER replanning in response to execution-time resource usage updates. In the CNSR domain, the data collected during the sample activities is compressed and then stored in the data buffer of the lander. This data is uplinked to the orbiting spacecraft at a later time. The planner uses estimates of the amount of data compression to plan when uplink activities are necessary. Because the compression algorithms are content dependent, these estimates may significantly deviate from actual achieved compression.

In this example, the actual data generated by the second sample activity is greater than expected because the compression achieved is less than originally estimated. The planner realizes that it will not have sufficient buffer memory to perform the third sample activity. This results in an over-subscription of the data buffer depletable resource. The planner knows that such a conflict can be repaired by: 1) removing activities that contribute to resource usage or 2) adding an activity that renews the resource. In this case these two options correspond to deleting the third sample activity or adding an uplink activity. (The uplink activity renews the buffer resource by uplinking data to the orbiter.) The planner resolves this conflict by adding an uplink activity after the second sample activity, freeing memory for the third sample activity (see Figure 5).

3.2   Planetary Rover Operations Testbed

The CASPER system has been tested on a model and simulation for the Rocky-7 prototype Mars rover for long-range planetary science gathering [10,11]. The model consists of 18 shared resources, 12 state variables and 32 activity types. Resources and states include 3 digital cameras (at the front, rear, and on a mast), a deployable mast, a shovel, a spectrometer, solar arrays, batteries, and a RAM buffer. There are five activity groups that correspond to different types of science experiments: one for collecting spectrometer readings, and four for taking images at a location (a front image, a rear image, a panorama using the mast, and a close-up image using the mast). Rover problems are sized by the number of hours of daylight (all operations require illuminated solar arrays). A series of science goals are generated corresponding to the number of hours of daylight, and the parameters for the goals are randomly generated (such as target locations). For each additional hour of daylight, ten additional goals are added spread over two new locations. Repair heuristics include traveling salesman heuristics, which attempt to order the rover traversals such that the total distance traveled is minimized.

Again, to show the richness of the domain, we describe the random variables included in the rover simulation.

- Traverse speed is modeled as $\sim N(0.016, 0.0016)$ meters per second. This is intended to model unpredictable traverse speeds due to terrain (elevation, traction) effects or due to obstacle avoidance.
- Power usage of a traverse is modeled as $\sim N(4, 0.4)$ watts to model unpredictable power draw from the wheel motors due to terrain (elevation, traction) effects or due to obstacle avoidance.
- Duration of localization activity is modeled as $\sim N(120, 12)$ seconds to model the variability in the amount of time taken to deploy the mast (variances in mechanical actuation) as well as the variable amount of time to acquire images and run the localization algorithm.
- Power usage of localization activity is modeled as $\sim N(1, 0.1)$ watts to account for variances in mast movements and camera pointing required for localization.
- Rate of data transmission for data transfer activity is modeled as $\sim N(10, 1)$ bytes per second. This is intended to model communications errors and retransmissions that may be required for communications from the rover to a lander or orbiter.

- Power usage for data transfer activity is modeled as ~N(1, 0.1) watts - data transfer consists of transmission, waiting for acknowledgements of receipt of data, and retransmission. Transmission and retransmissions are power intensive, during the waits power consumption is reduced.
- Power usage of the spectrometer activity is modeled as ~N(1, 0.1) watts. The spectrometer activity includes deployment of the spectrometer to the surface of a rock, gathering data, and stowing the instrument. The deployment and stowage require movements of the arm which are dependent on the exact geometries of the surface being sample. These movements and feedback often cause variations in the amount of time and power draw for this activity.
- The amount of data generated (memory usage) for the spectrometer activity is modeled as ~N(0.2, 0.02) bytes per second to account for the effects of content dependent compression.
- Position estimation error rate of increase is modeled as ~N(0.02, 0.002). The position error is the rover's model of how well it knows its position from wheel sensors, etc. This increases at varying rates based on terrain variations as the rover moves.
- Actual rover movements are modeled as a ~N(0, 0.0025) meters per second wander away from the commanded path (0.5 meters away from target location is too far for a science experiment). The actual location vector state variable is used to determine actual distances to feed traverse activity parameters (such as travel time). Variability here is caused by terrain effects and obstacle avoidance as well as position estimation error.

4   Empirical Results

In our tests we compare two different local search methods.

1. The CASPER approach - which uses the current plan as an initial seed and performs local, iterative repair to attempt to make the plan consistent with the current state.
2. Batch replanning on failure - in this approach after an update the plan cleared and then local, iterative repair is invoked to construct a plan to achieve the top-level goals from scratch. This differs from the CASPER algorithm in that the seed plan for local search is the null plan (e.g., no actions).

In our setup, CASPER was running on a Sun workstation Ultra 60 with a 359 MHz processor with 1.1 GB Memory. The experiment consisted of 200 runs each (*batch* and *repair*) on randomly generated problems. During each run, the simulator updates the plan an average of 18,000 times (most of these are battery power level updates). On average, around 100 updates result in conflicts that should be handled by the planner/scheduler.

In order gain insight into the effectiveness of the local search in replanning, we examine the number of plan operations and CPU time to either repair the current plan or to construct a new plan to reflect the real-time execution feedback. Table 1 shows the average number of CPU seconds and plan modifications required to repair the existing old plan and construct a new plan from scratch. This data clearly shows that it is more efficient to re-use the old plan than to construct a plan from scratch.

Table 1. Summary of Data Comparing Average # of Plan Iterations and CPU Seconds required to Repair Old Plan vs. Construct New Plan from Scratch.

Domain	Average		Average	
Domain	Planning Iterations		CPU Seconds	
	Repair Old Plan	Batch Plan	Repair Old Plan	Batch Plan
CNSR	19.73	67.60	2.55	8.10
Rover	15.9	33.6	1.80	3.43

Table 2 shows, for the CNSR simulations, a histogram plot which indicates the frequency with which problems from feedback required a given number of plan operations to repair (i.e. the height of the column on the Y axis indicates the frequency of problems requiring the number of repair operations indicated on the X axis). Table 3 shows a similar plot for the CPU Time for the CNSR scenario. These plots indicate that a large number of solutions to the replanning problem lie in the neighborhood of the old plan. This validates our hypothesis that if the Δgoals and Δstate are small, that the Δplan should also be small. Additionally the histograms clearly show that replanning from the old plan is more efficient than batch planning from scratch.

We also briefly present results in a rover operations domain (space constraints preclude a more lengthy presentation of these results). Tables 4 and 5 show the histogram plots for the plan operations and CPU time required to repair the old

Table 2. CNSR – Number of planning operations

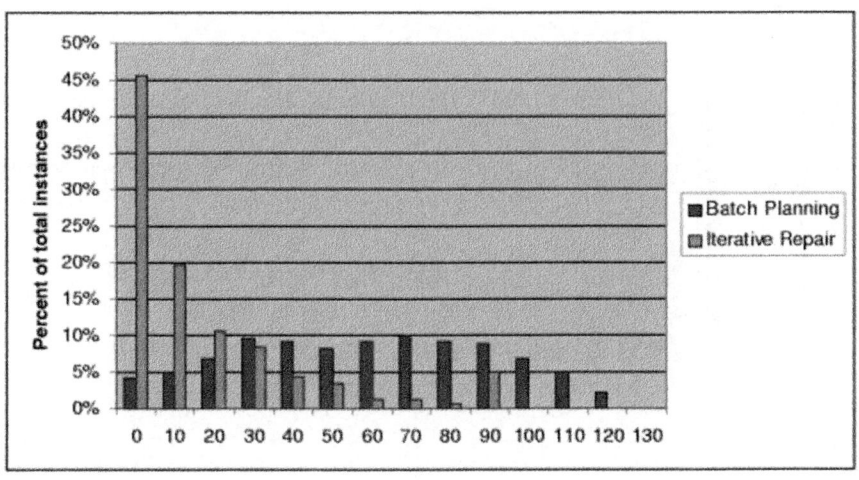

Table 3. CNSR – Time, In seconds

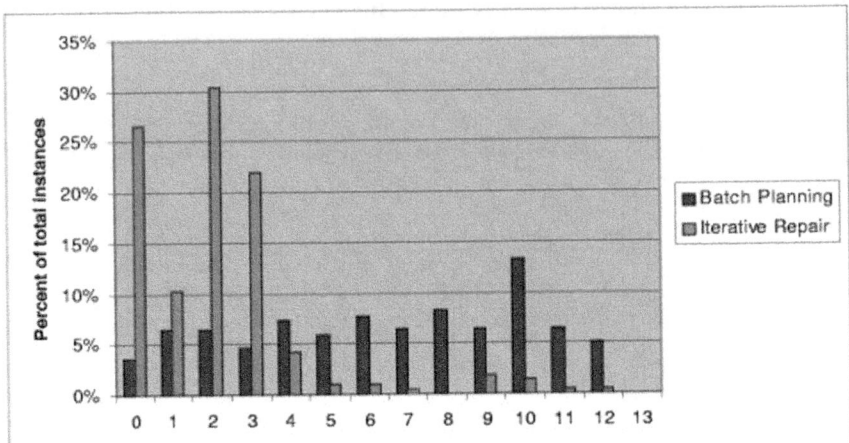

plan vs. constructing a new plan from scratch. Again the data indicates that not only are a large number of solutions quite close to the old plan but that re-using the old plan is significantly more efficient than replanning from scratch.

Note that while the absolute differences might not seem large, the relative differences are quite significant. The CNSR domain shows a larger than 3x speedup and the Rover domain shows a speedup of almost 2x. These factors become critical when the running on a slower flight processor, which can typically increase CPU times by several orders of magnitude. The data on plan operations was

Table 4. Rover – Number of planning operations

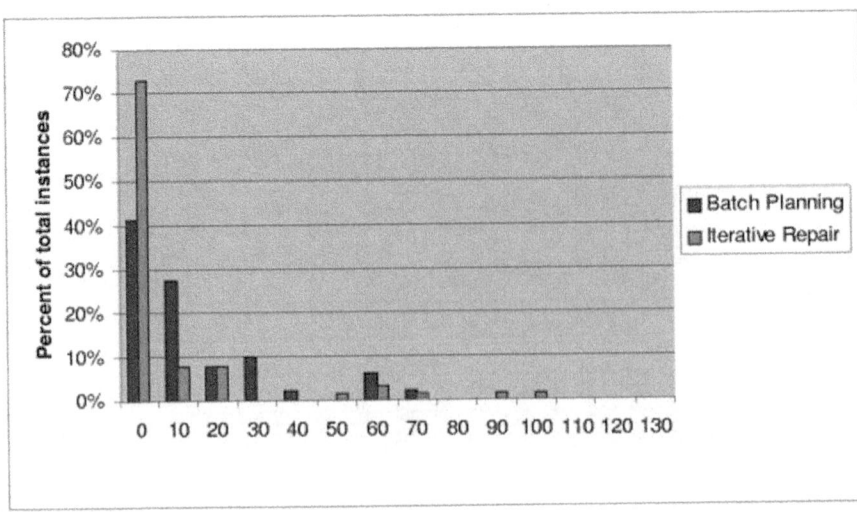

Table 5. Rover – Time, In seconds

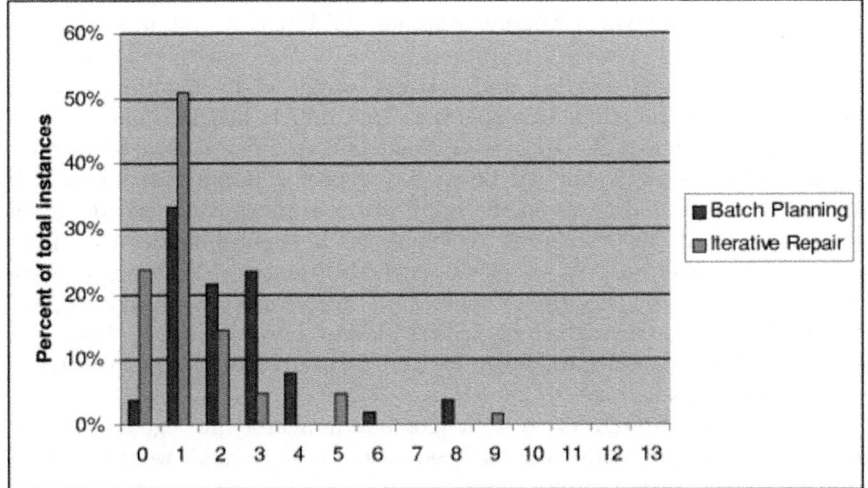

provided to give some measure that is not CPU dependent. However, plan operations are specific to the chosen planner and CASPER's plan operations can vary significantly, with some requiring more time than others.

5   Related Work

The question of repairability that this paper addresses is strongly linked to the notion of supermodels [12]. In supermodels, they examine the problem of finding a (m,n) SAT model which after having m bits flipped, can be made consistent by flipping a different n bits. They show that for specific classes of problems this can be reduced to a SAT problem. In contrast, we are interested in finding a plan that after being perturbed by real-world feedback (corresponding to the m bits flipped above) that using a bounded amount of computation (n bits flipped in response above) can be repaired to be consistent (be made a consistent SAT model). Because our perturbation space is much more rich (stochastic elements for states, resource usage, and activity duration) and our plan repair space more rich (add, move, delete, abstract activities) the problems are alike only at the most abstract level. However, the general approach of trying to generate robust plans is exactly the problem of interest and the study in this paper is aimed at evaluating the ability of local search to repair plans.

The high-speed local search techniques used in our continuous planner prototype are an evolution of those developed for the DCAPS system [1] that has proven robust in actual applications. In terms of related work, iterative algorithms have been applied to a wide range of computer science problems such as traveling salesman [13] as well as Artificial Intelligence Planning [4,14,15,16, 17]. Iterative repair algorithms have also been used for a number of scheduling systems. The GERRY system [4] uses iterative repair with a global evaluation function and simulated annealing to schedule space shuttle ground processing activities. The Operations Mission Planner (OMP) [14] system used iterative repair in combination with a historical model of the scheduler actions (called chronologies) to avoid cycling and getting caught in local minima. Work by Johnston and Minton [18] shows how the min-conflicts heuristic can be used not only for scheduling but also for a wide range of constraint satisfaction problems.

The OPIS system [6] can also be viewed as performing iterative repair. However, OPIS is more informed in the application of its repair methods in that it applies a set of analysis measures to classify the bottleneck before selecting a repair method. Excalibur [19] represents a general framework for using constraints to unify planning and scheduling constraints, uncertainty, and knowledge. This framework is consistent with the CASPER design, however in this paper we have focused on a lower-level. Specifically, we have focused on re-using the current plan using iterative repair.

Work on the PRODIGY system [20] has indicated how goals may be altered due to environmental changes/feedback. These changes would be modeled in our framework via task abstraction/retraction and decomposition for potentially failing activities. Other PRODIGY work [21] has focused on determining which elements of the world state need to be monitored because they affect plan appropriateness. In our approach we have not encountered this bottleneck, our fast state projection techniques enable us to detect relevant changes by noting the introduction of conflicts into the plan.

Work on CPEF (Continuous Planning and Execution Framework) [22] uses PRS, AP, and SIPE-2, also represents a similar framework to integrating planning and execution. CPEF and CASPER differ in a number of ways. First, CPEF attempts to cull out key aspects of the world to monitor (as is necessary in general open-world domains). They also suggest the use of iterative repair (they use the term conservative repairs). And their taxonomy of failure types is very similar to ours in terms of action failure and re-expansion of task networks (re-decomposition). However, in this paper we have focused on lower level issues in synchronization and timing.

6   Discussion and Conclusions

There are a great many areas for future work, the work presented in this paper is just beginning to analyze local search for replanning. Attempting to characterize better the cases under which problems are nearly decomposable would be extremely useful, as this property is strongly related to whether or not small changes in goal and state would require major changes to the plan. Understanding how abstraction could assist in making large plan changes more tractable is also important. Relating this work to other (non-planning) work in using old invalid solutions as a guide to solve new (slightly modified) problems is also important. For example, dynamic constraint satisfaction has investigated this area. Also, solution density and the phase transition are related to the notion of tractable local search for replanning - better understanding this relationship would be ideal. Finally, one interpretation of the results is that our planner heuristics are more well suited for repair (modification) of an existing plan than for construction of a plan from scratch. This is almost certainly true - but in general we believe that it is easier to construct good heuristics for repairing a plan than for from scratch plan construction. Benchmarking against other from scratch planning methods would be useful.

This paper has described an empirical evaluation of a local search approach to integrating planning and execution for spacecraft control and operations. In this empirical study we investigated the hypothesis that small perturbations in execution of a plan would be resolvable in an efficient fashion by local search. Empirical evidence from two space mission simulations supports the use of local search for this type of problem.

Acknowledgements. This work was performed by the Jet Propulsion Laboratory, California Institute of Technology, under contract with the National Aeronautics and Space Administration.

The authors would like to thank the anonymous reviewers for their many insightful comments that improved the readability of the paper and indicated many excellent areas for future work.

References

1. S. Chien, G. Rabideau, J. Willis, and T. Mann, "Automating Planning and Scheduling of Shuttle Payload Operations," *Artificial Intelligence Journal*, 114 (1999) 239-255.
2. NASA Ames & JPL, Remote Agent Experiment Web Page, http://rax.arc.nasa.gov/, 1999.
3. S. Minton "Automatically Configuring Constraint Satisfaction Programs: A Case Study." *Constraints* 1:1(7-43).1996.
4. M. Zweben, B. Daun, E. Davis, and M. Deale, "Scheduling and Rescheduling with Iterative Repair," in *Intelligent Scheduling*, Morgan Kaufman, San Francisco, 1994.
5. S. Chien, G. Rabideau, R. Knight, R. Sherwood, B. Engelhardt, D. Mutz, T. Estlin, B. Smith, F. Fisher, T. Barrett, G. Stebbins, D. Tran , "ASPEN - Automating Space Mission Operations using Automated Planning and Scheduling," Space Operations 2000, Toulouse, France, June 2000.
6. S. Smith, "OPIS: An Architecture and Methodology for Reactive Scheduling," in *Intelligent Scheduling*, Morgan Kaufman, 1994.
7. S. Chien, R. Knight, A. Stechert, R. Sherwood, and G. Rabideau, "Using Iterative Repair to Improve Responsiveness of Planning and Scheduling," *Proc. 5th International Conference on Artificial Intelligence Planning and Scheduling*, Breckenridge, CO, April 2000.
8. G. Rabideau, R. Knight, S. Chien, A. Fukunaga, A. Govindjee, "Iterative Repair Planning for Spacecraft Operations in the ASPEN System," Int Symp on Artificial Intelligence Robotics and Aut. in Space (ISAIRAS), Noordwijk, The Netherlands, June 1999.
9. Jonsson, P. Morris, N. Muscettola, K. Rajan and B. Smith, "Planning in Interplanetary Space: Theory and Practice," *Proceedings of the Fifth International Conference on Artificial Intelligence Planning Systems.* Breckenridge, CO. April, 2000.

10. R. Volpe, J. Balaram, T. Ohm, and R. Ivlev, "Rocky 7: A Next Generation Mars Rover Prototype," *Journal of Advanced Robotics*, 11(4), December 1997.
11. S. Hayati, and R. Arvidson, "Long Range Science Rover (Rocky 7) Mojave Desert Field Tests," *Proceedings of the 1997 International Symposium on Artificial Intelligence, Robotics and Automation in Space*, Tokyo, Japan, July 1997.
12. M. Ginsberg and A. Parkes, "Supermodels and Robustness," Proceedins of AAAI-98.
13. S. Lin and B. Kernighan, "An Effective Heuristic for the Traveling Salesman Problem," *Operations Research Vol. 21*, 1973.
14. E. Biefeld and L. Cooper, "Bottleneck Identification Using Process Chronologies," *Proceedings of the 1991 International Joint Conference on Artificial Intelligence*, Sydney, Australia, 1991.
15. S. Chien and G. DeJong, "Constructing Simplified Plans via Truth Criteria Approximation," *Proceedings of the Second International Conference on Artificial Intelligence Planning Systems*, Chicago, IL, June 1994, pp. 19-24.
16. K. Hammond, "Case-based Planning: Viewing Planning as a Memory Task," Academic Press, San Diego, 1989.
17. G. Sussman, "A Computational Model of Skill Acquisition," Technical Report, MIT Artificial Intelligence Laboratory, 1973.
18. M. Johnston and S. Minton, "Analyzing a Heuristic Strategy for Constraint Satisfaction and Scheduling," in *Intelligent Scheduling*, Morgan Kaufman, San Francisco, 1994.
19. Nareyek, "A Planning Model for Agents in Dynamic and Unicertain Real-Time Environments," in Integrating Planning, Scheduling, and Execution in Dynamic and Uncertain Environments, AIPS98 Workshop, AAAI Technical Report WS-98092.
20. M. Cox & M. Veloso, "Goal Transformation in Continuous Pannning," in Proceedings of the AAAI Fall Symposium on Distributed Continual Planning, 1998.
21. M. Veloso, M. Pollack, M. Cox, "Rationale-based monitoring for planning in dynamic environments," Proceedings Artificial Intelligence Planning Systems Conference, Pittsburgh, PA, 1998.
22. K. Myers, "Towards a Framework for Continuous Planning and Execution", in Proceedings of the AAAI Fall Symposium on Distributed Continual Planning, 1998.

Board-Laying Techniques Improve Local Search in Mixed Planning and Scheduling

Russell Knight, Gregg Rabideau, and Steve Chien

Jet Propulsion Laboratory, California Institute of Technology
4800 Oak Grove Drive, Pasadena, CA 91109-8099
{knight, rabideau, chien}@aig.jpl.nasa.gov

Abstract. When searching the space of possible plans for combined planning and scheduling problems we often reach a local maximum and find it difficult to make further progress. To help move out of a local maximum, we can often make large steps in the search space by aggregating constraints. Our techniques improve the performance of our planner/scheduler on real problems.

1 Introduction

This paper describes an approach that can improve search in the plan space. This is achieved by "jumping" within the search space by using an aggregation technique. Additional costs of the technique include the overhead involved in reasoning about the aggregation. Previous work in the form of macro operators and clustering addresses the problem of deciding on an appropriate collection of activities [1] heretofore referred as an aggregation. Our aggregate reasoning techniques [2] are applicable to any collection of activities with interacting constraints, regardless of how the collection was formed. We apply a simple heuristic technique to construct the aggregation and then reason about the aggregation as a single unit. Empirical data show that our aggregate reasoning techniques facilitate local, repair search. While we believe that these techniques would also improve other types of search, this is still an open area of research.

1.1 Board-Laying: An Analogy of Aggregation

Consider searching the space of plans, where a plan consists of a collection of activities possibly related to one another with temporal constraints as in Figure 1. Consider that each activity also has resource and state constraints or *reservations* (Figure 2).

Now consider a repair search where a *conflict* is a violation of either a temporal constraint or a reservation. If we start with a plan that contains conflicts of temporal constraints and reservations, we might decide to repair the temporal constraint conflicts and then the reservation conflicts (e.g., temporal constraints alone could be resolved using algorithms from [3]). When we have repaired all of the temporal con-

A. Nareyek (Ed.): Local Search for Planning and Scheduling, LNAI 2148, pp. 95–107, 2001.
© Springer-Verlag Berlin Heidelberg 2001

straint conflicts, we have climbed a hill to a certain point, where hill climbing is seen as reducing conflicts. (Note: this is the dual of the "gradient descent" analogy in which valleys are more optimal than hilltops.) If we have no other temporal constraint conflicts then we would attempt to resolve the reservation conflicts, but it may be that resolving a reservation conflict implies causing temporal constraint conflicts. In this case, we have reached a local maximum—in the sense of hill climbing, we have reached a hilltop, as in Figure 3.

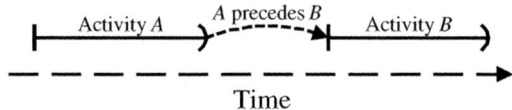

Fig. 1. A plan consisting of two activities related to each other via a temporal constraint.

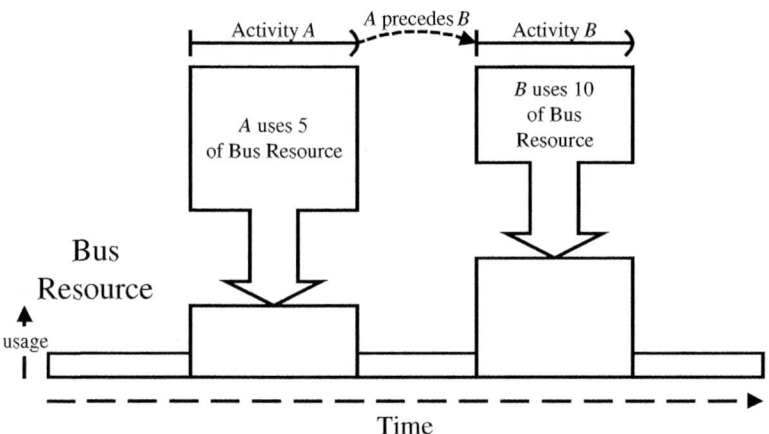

Fig. 2. Reservations of activities on a shared resource

One option is to simply descend the hill and solve the reservation conflicts, and then re-solve the temporal constraint conflicts. But most real domains we face have too many options. In an exhaustive search, this leads to large amounts of backtracking and an impossibly long search. In a local search, we may find ourselves descending and ascending the same hill, making no progress toward a global solution, and relying more heavily on locally sub-optimal, stochastic moves.

What we really want is to move from one hilltop to other local hilltops without descending. This is advantageous in that we know that we are not ascending the same hill and that we continue the search "up". One way of enabling this sort of operation is to reason about collections of activities instead of individual activities. This allows us to avoid some of the descending associated with reasoning about individual activities.

In this sense, aggregation is equivalent to laying boards from one hilltop to nearby hilltops, as in Figure 4.

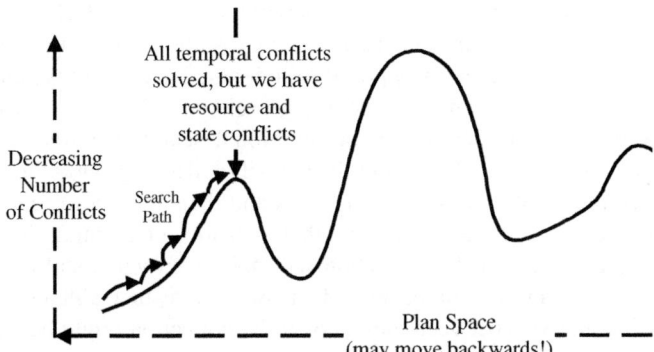

Fig. 3. Reaching a hilltop in the search of the plan space

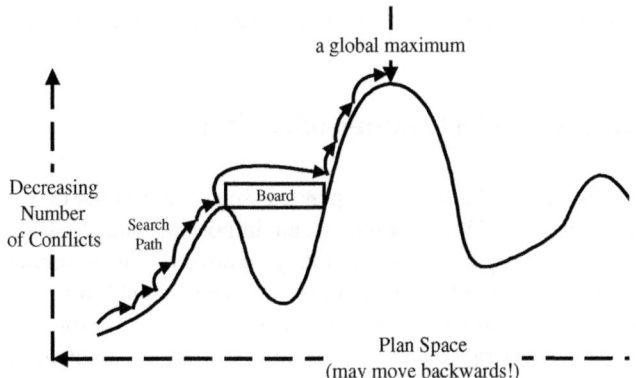

Fig. 4. Efficiently achieving a less-conflicting plan using board laying

Unfortunately, board-laying is not free. It incurs the addition cost of constructing the board. In other words, we must reason about the interactions between shared states and resources among the activities in the aggregation in order to make appropriate aggregate moves.

Our criterion for inclusion into an aggregation is motivated by the structure of problems presented by modelers to our planner/scheduler: ASPEN [4,5,6]. Often, a complex activity is modeled as a collection of simpler activities related to each other via temporal constraints. Therefore, we first choose an activity that has a reservation violation, and then we simply gather associated activities by calculating the connected component of the original activity in the temporal constraint network. This takes advantage of the intent of the modelers in that we assume that required activities to satisfy temporal constraints will need to be considered while satisfying resource and state constraints. Obviously, if this is not the case our aggregation criterion will not produce helpful collections. Therefore, we make no claims as to the appropriateness of our aggregation technique—we only claim that it is justifiable given the models actually produced for domains without respect to any particular solution technique.

To summarize the concept and issues of aggregation, we believe that reasoning about collections of activities (as opposed to individual activities) is helpful in finding a solution to a combined planning/scheduling problem. We believe this is due to the increased complexity introduced by the additional search required when reasoning about activities in isolation. To avoid this, we will collect a group of activities that are related to each other with temporal constraints and then attempt to schedule the entire collection at once. In a sense, we assume that decisions concerning the temporal constraints among members of the collection will hold even when shifted "lock-step" in the schedule either forward or backward in time. Note that although research concerning flexible intervals between members of the collection might be fruitful, it is left as future work.

Given that we have chosen a collection of activities, does our technique: 1) improve on search efficiency and 2) improve search on any reasonable domain? Our empirical analysis compares our technique with a technique that performs no aggregation (answering 1), and we use domains from space-exploration (answering 2).

2   Reasoning about Collections of Activities

Consider the task of scheduling interdependent sets of activities in a combined scheduling/planning problem. This problem is an important aspect of solving combined planning and scheduling problems. In many approaches to combined planning and scheduling, one alternates between finding activities to satisfy pre- and post-conditions (planning) and finding temporal assignments and resources for those activities (scheduling). Complex activity placement is also an important component of many scheduling problems, as finding temporal assignments for complex activities can be computationally challenging.

This work advances the approach of moving collections of activities whose temporal relationships among them are fixed. We proceed by describing our motivation, defining the problem, and describing the solution. Finally, we present empirical evidence in favor of our technique.

We assume that we have already gathered activities into an aggregate, as described earlier. We explicitly represent interactions between the activities, or more specifically, interactions between the constraints on shared states and resources.

For example, consider a pair of activities that affect a battery (see Figure 5). The first activity a_1 uses 10 amp-minutes, while the second activity a_2 restores 10 amp-minutes. If we schedule a_1 individually, we find no intervals that will not cause an over-subscription of the battery, because activity a_3 has already fully depleted the battery by the end of the current schedule. But, if we schedule these together, we find placements that are valid. The positive effect a_2 has on the schedule makes up for a_1's usage. Our aggregate reasoning techniques handle this sort of constraint-interaction for shared states and resources.

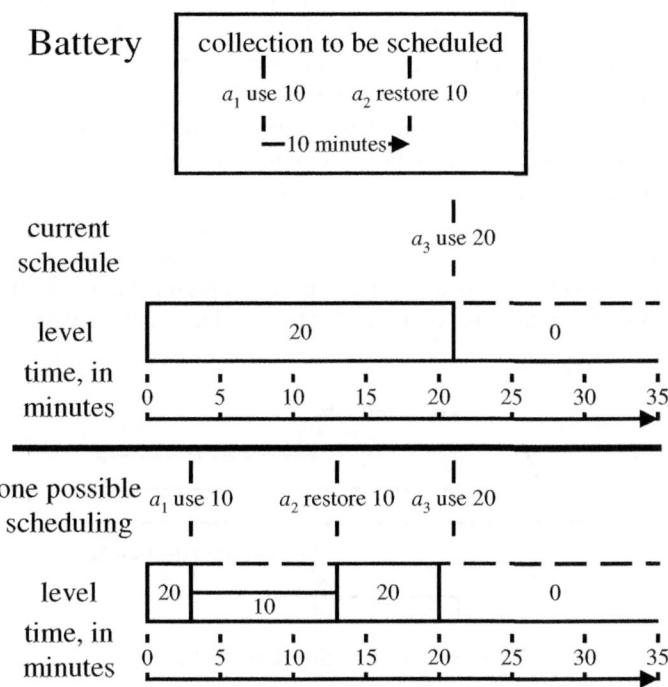

Fig. 5. Battery interaction example

More formally, the problem of computing valid intervals is as follows:

Given: a schedule (S) and a collection of activities (Q) to be scheduled whose temporal relationships among themselves are fixed

Compute: the interval set (I) representing assignments of the start-time of the earliest activity in the collection that violate no constraints

And our approach to computing I is as follows:
1) gather all reservations of activities in Q,
2) partition these according to timeline
3) compute the valid intervals for each partition (P),
4) translate and intersect the valid intervals.

We focus on step 3. We know that if the reservations in P do not interact, then we can compute valid intervals for P by computing those for each individual reservation in P. This is accomplished by first computing each set of intervals for each individual reservation in P. We then translate these intervals by the difference between the start-time of the reservation and the earliest start-time in P. Finally, the resultant intervals are intersected to compute I.

If the reservations in P *do* interact, one way to make step 3 tractable is to transform P into a set of non-interacting reservations P'. Because the reservations in P' do not interact, we can compute valid intervals using simple translation and intersection of the valid intervals for each individual reservation in P'.

The transformation of P into P' depends on the type of reservations involved. Generating P' for reservations on depletable timelines, as in the battery example, is shown in Figure 6. First, we create a new set of local reservations for all but the last reservation in P that represent the changes in the value of the timeline along the temporal extent of the of the reservations in P. Finally, we add a single downstream reservation at the end of the collection of local reservations that represents the overall effect of the reservations in P. Note that the semantics for reservations in P' may differ from those in P.

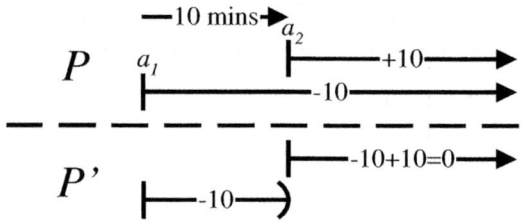

Fig. 6. Transforming interacting depletable reservations

Similar methods are used to compute P' for other types of timelines. Once we have our complete set of reservations, valid intervals for each can be computed in time that is proportional to the number of reservations already scheduled. The number of reservations in P' is a constant factor more than those in P, making the overall asymptotic complexity of our technique equivalent to a computation that ignores interactions, or roughly proportional to $|P| \cdot |A|$ where A is the number of reservations already scheduled.

3   Local Search

To solve our planning/scheduling problems, we use the ASPEN system using an "iterative repair" search algorithm [4,7], that classifies conflicts and attacks them each individually. Conflicts occur when a plan constraint has been violated; this constraint could be temporal or involve a resource or state timeline. Conflicts are resolved by performing one or more plan modifications such as moving, adding, or deleting activities. Heuristics can be used to help make selections among alternatives. The iterative repair algorithm continues until no conflicts remain in the plan, or until the computational resource bounds have been exceeded.

When moving activities, the planner/scheduler entertains least-conflicting locations in the plan. This is when aggregate reasoning can be applied. The collection is determined by the selected activity, which is typically chosen at random. For all temporal constraints for which the selected activity is the source of the constraint, we include all activities participating in the constraints. It is worth noting that the selected activity may participate in other temporal constraints for which it is not the source. These constraints are not considered and may become violated. However, this also provides a mechanism for changing the relative position of activities to satisfy temporal constraints between them. Once the collection is established, aggregate reasoning is used to evaluate interactions between resource and state constraints of activities in the collection. This is done to compute locations for the collection that will not violate these constraints. These non-conflicting locations are used in a heuristic for selecting the new start time for the selected activity.

There are several important characteristics of our local search algorithm. First, the search is non-systematic and each iteration attempts to repair a conflict with no knowledge of previous iterations. In other words, no history of the search is maintained. In order to ensure that we cover the search space, decisions are made stochastically whenever heuristics make no preference between choices. While we believe this sort of local, stochastic search is an effective means of achieving good performance on real problems, we do not believe it is a requirement for aggregate reasoning.

Another important characteristic of our search is that it can be categorized as an early-commitment approach. As opposed to least-commitment search, early-commitment search commits to a single choice for a decision rather than maintain all possible non-conflicting choices. This choice may result in new conflicts (either immediately or after further planning) and it is left to the search to repair the new conflicts on subsequent iterations. While algorithms exist for maintaining least-commitment information (e.g., temporal constraint networks) and hence reducing search, we believe these algorithms add an overhead that is too costly to be used at every step of the search. This also appears to be true for aggregate reasoning algorithms. Our current implementation of aggregate reasoning is very efficient but only applicable to activities and timelines with committed values (e.g., start times, resource usages). We believe that our general technique is applicable to least-commitment planning, but at a greater computational expense (while still being polynomial). Future research should address this issue.

4 Empirical Evaluation

In our empirical analysis we use four models (and corresponding problem set generators): 1) the EO1 spacecraft operations domain, 2) the Rocky-7 Mars rover operations domain, 3) the DATA-CHASER shuttle payload operations domain, and 4) the New Millennium Space Technology Four landed operations domain.

Within each model and corresponding problem set, we generate random problems that include a background set of fixed activities and a number of movable activity groups. The activity groups are placed randomly. The goal is to minimize the number of conflicts in the schedule by performing planning and scheduling operations. The main scheduling operation is to move activities to new start times. In the *control* trials the planner/scheduler always moves one activity at a time. We dub this the *atomic* technique. In the *experiment* trials the planner/scheduler is using our aggregate reasoning (i.e., *board-laying*) to compute valid placements for collections of activities and the collection is moved as a whole. In all cases for each domain, both trials are using the same set of heuristics at all other choice-points (e.g., selection of a conflict or activity group to attempt to repair, where to place within computed valid intervals, etc.). Note that simple (atomic) operations are available in all domains.

We now briefly describe each domain including information on the types of activities and resources modeled, what the activity groups are, and how they are interdependent.

EO1 Domain

The EO1 domain models the operations of the New Millennium Earth Observer 1 operations for a two-day horizon [6]. It consists of 14 resources, 10 state variables and total of 38 different activity types. Several activity groups correspond to activities necessary to perform different types of instrument observations and calibrations. The activity groups range in size from 23 to 56 activities, many of which have interactions. For example, taking an image of the earth requires fixing the solar array drive to avoid blurred images. The high-level observation activity group includes both commands to fix the SAD and take the image.

Each EO1 problem instance includes a randomly generated, fixed profile that represents typical weather and instrument pattern. Each problem also includes 8 randomly placed instrument requests for observations and calibrations.

Rocky-7 Domain

The Rocky-7 Mars rover domain models operations of a prototype rover for a typical Martian day [8]. It consists of 14 shared resources, 7 state variables and 25 activity types. Resources and states include cameras (front, rear, mast), mast, shovel, spectrometer, solar array, battery, and RAM. There are four activity groups that correspond to different types of science experiments: imaging a target, digging at a location, collecting a spectrometer reading from target, and taking a panoramic image from a location. Activity group size ranges from 8 to 17 activities. Members in activity groups have positive resource interactions, e.g. opening the aperture for the camera enables subsequently taking a picture. Activity groups also have negative interactions, e.g. several member activities using the onboard buffer. Rover problems are constructed by generating four experiments and randomly generating parameters for the experiments (such as target locations).

New Millennium Space Technology Four Landed Operations Domain

The ST4 domain models the landed operations of a spacecraft designed to land on a comet and return a sample to earth. This model has 6 shared resources, 6 state variables, and 22 activity types. Resources and states include battery level, bus power, communications, orbiter-in-view, drill location, drill state, oven states for a primary and backup oven state, camera state, and RAM. There are two activity groups that correspond to different types of experiments: 1) mining and analyzing a sample, 2) taking a picture. Activity group sizes range from 5 to 10. As in the rover domain, activities interact positively and negatively.

Each ST4 problem instance includes a randomly generated, fixed profile that represents communications visibility to the orbiting spacecraft. Each problem also includes five mining and two picture experiments (each randomly placed.)

DATA-CHASER Domain

The DCAPS domain models operations of a shuttle science payload that flew onboard Space Shuttle Flight STS-85 in August, 1997 [9]. It consists of 19 shared resources, 25 state variables, and 70 activity types. Resources and states include shuttle orientation, contamination state, 3 scientific instruments (doors, relays, heaters, etc.), several RAM buffers, tape storage, power (for all instruments/devices), and downlink availability. There is one type of activity group corresponding to one experiment for each of the 3 scientific instruments. This activity group consists of 23 activities. As with the other domains, activities in this activity group interact positively and negatively.

Each DCAPS problem instance includes a randomly generated, fixed profile that represents shuttle orientation and contamination state. The number of randomly placed experiments ranges from 2 to 20 based on the fixed profile for the given problem instance.

Comparison

For each domain, we run 20 random problems for both the control (atomic) and the experimental techniques. Using the *atomic* technique, the problems are intractable within reasonable time bounds. We postulate that this is because the distance in terms of sub-optimal moves from one local optima to the next is $O(n)$ and the space to be searched is $O(m^n)$ where n is the number of activities in a movable collection and m is the number of possible locations given by a calculation of legal intervals for an individual activity. For example, in the EO1 domain, n ranges from 23 to 56; in the Rover domain, n ranges from 8 to 17.

However, we discover that our board-laying technique fairs somewhat better (see Table 1). Also, we see that as a function of time, our board laying technique outperforms the atomic technique for each domain (see Figure 7, Figure 8, Figure 9 and Figure 10).

Table 1. Number of Problems Successfully Planned

	EO1	Rocky 7	DCAPS	ST4	Total
Board laying	149/400	390/400	387/400	243/400	1169/1600

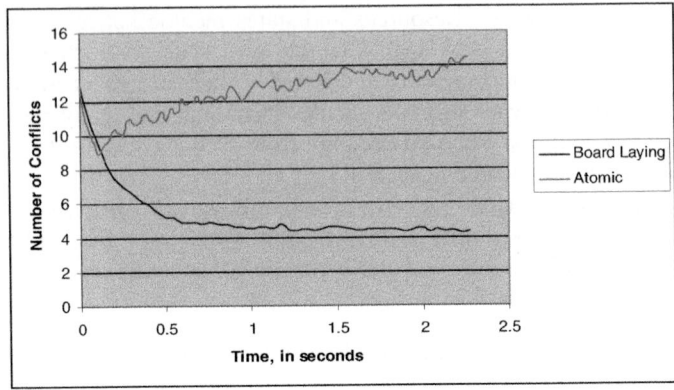

Fig. 7. Conflict Reduction for EO1

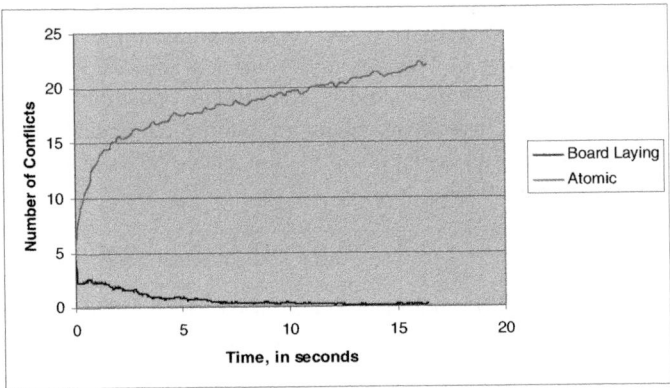

Fig. 8. Conflict Reduction for Rocky 7

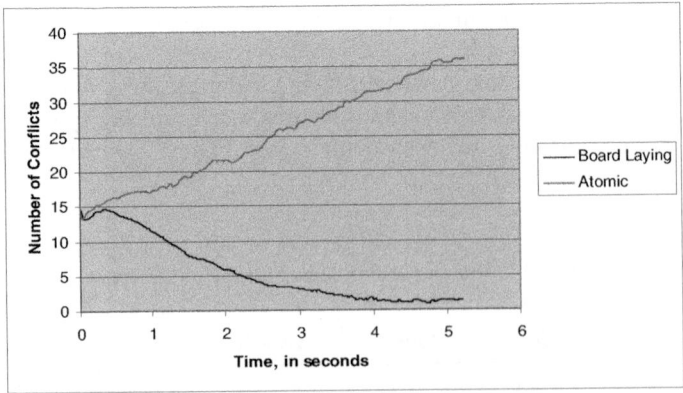

Fig. 9. Conflict Reduction for DCAPS

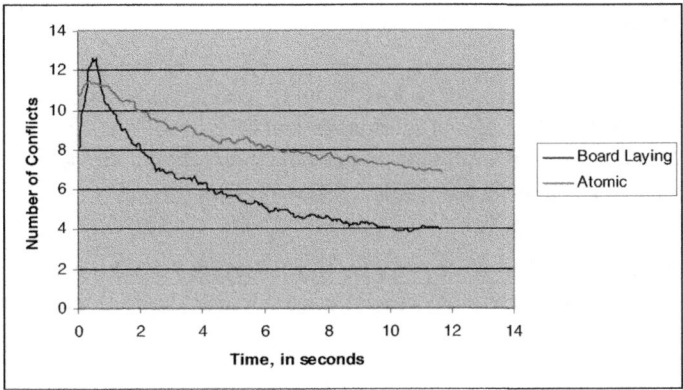

Fig. 10. Conflict Reduction for ST4

5 Related Work

There are a number of related systems that perform both planning and scheduling. IxTeT [10] uses least-commitment approach to sharable resources that does not fix timepoints for its resource and state usages.

HSTS [11] enforces a total order on timepoints that affect common shared states and resources, allowing more temporal flexibility. We believe that our technique is applicable in this case at a greater computational expense (while still being polynomial), and future research should address this issue.

Both IxTeT and HSTS are less committed representations than our grounded time representation and this flexibility incurs a greater computational expense to detect and/or resolve conflicts.

O-PLAN [12,13] also deals with state and resource constraints. O-PLAN's resource reasoning uses optimistic and pessimistic resource bounds to efficiently guide its resource analysis when times are not yet grounded. Like ASPEN, O-PLAN also allows multiple constraint managers which would enable it to perform general reasoning when times are unconstrained and more efficient reasoning in the case where all timepoints are grounded.

SIPE-2 [14] handles depletable/non-depletable resource and state constraints as planning variables using constraint posting and reasons at the same level of commitment as IxTeT.

In [15], constraint-posting techniques are applied to satisfy multi-capacitated resource problems at the same level of commitment. Depletable/non-depletable resource constraints are easily transformed to multi-capacitated resource constraints. None of these systems generally consider aggregate operations in their search space.

There also exists a large body of work concerning the flattening of search spaces. Future work should include comparisons with these techniques, an example of which is Tabu search [16].

6 Conclusion

This paper has described the use of board laying techniques to improve the efficiency of planning and scheduling sets of interdependent activities. We show empirically that our board laying search method outperforms the alternative approach of using singleton operations on problems from space exploration domains.

Acknowledgements. This work was performed at the Jet Propulsion Laboratory, California Institute of Technology, under contract with the National Aeronautics and Space Administration.

References

1.   Estlin, T., Chien, S., and Wang, X. 1997. "An Argument for an Integrated Hierarchical Task Network and Operator-based Approach to Planning," in *Recent Advances in AI Planning*, S. Steel and R. Alami (eds.), Notes in Artificial Intelligence, Springer-Verlag, 1997, pp. 182–194.
2.   Knight, R., Rabideau, G., and Chien, S. 2000. "Computing Valid Intervals for Collections of Activities with Shared States and Resources," *Proceedings of the Fifth International Conference on Artificial Intelligence Planning and Scheduling*, 14-17 April 2000, pp. 339–346.
3.   Dechter, R., Meiri I., and Pearl J. 1991. "Temporal Constraint Networks," *Artificial Intelligence*, 49, 1991, pp 61–95.
4.   Rabideau, G., Knight R., Chien S., Fukunaga A., Govindjee A. 1999, "Iterative Repair Planning for Spacecraft Operations Using the ASPEN System," *iSAIRAS '99*, Noordwijk, The Netherlands, June 1999.
5.   Fukunaga, A., Rabideau, G., Chien, S., Yan, D. 1997. "Towards an Application Framework for Automated Planning and Scheduling," *Proc. of the 1997 International Symposium on Artificial Intelligence, Robotics and Automation for Space*, Tokyo, Japan, July 1997.
6.   Sherwood, R., Govindjee, A., Yan, D., Rabideau, G., Chien, S., Fukunaga, A. 1998. "Using ASPEN to Automate EO-1 Activity Planning," *Proc. of the 1998 IEEE Aerospace Conference*, Aspen, CO, March 1998.
7.   Zweben, M., Daun, B., Davis, E., and Deale, M., 1994. "Scheduling and Rescheduling with Iterative Repair," in *Intelligent Scheduling*, Morgan Kaufman, 1994.
8.   Rabideau, G., Chien, S., Backes, P., Chalfant, G., and Tso, K. 1999. "A Step Towards an Autonomous Planetary Rover," *Space Technology and Applications International Forum*, Albuquerque, NM, February 1999.
9.   S. Chien, G. Rabideau, J. Willis, and T. Mann, "Automating Planning and Scheduling of Shuttle Payload Operations," Artificial Intelligence Journal, 114 (1999) 239-255.
10. Laborie, P., Ghallab, M. 1995. "Planning with Sharable Resource Constraints," Proc. IJCAI-95, 1643–1649
11. Muscettola, N. 1993. "HSTS: Integrating Planning and Scheduling." *Intelligent Scheduling*. Morgan Kaufmann, March 1993.
12. Drabble, B., and Tate A. 1994. "Use of Optimistic and Pessimistic Resource Profiles to Inform Search in an Activity Based Planner," *Proc. AIPS94*.

13. Tate A., Drabble, B., and Dalton, J. 1996. "O-Plan: a Knowledge-based planner and its application to Logistics," *Advanced Planning Technology, Technological Achievements of the ARPA/RL Planning Initiative*, AAAI Press, 1996, pp. 259–266.
14. Wilkins, D., 1998. "Using the SIPE-2 Planning System: A Manual for Version 4.22." SRI International Artificial Intelligence Center, Menlo Park, CA, November 1998.
15. Cesta, Oddi S., and Smith S. 1998. "Profile-Based Algorithms to Solve Multiple Capacitated Metric Scheduling Problems." *Proc. AIPS98*, pp. 214–223.
16. Glover, F. 1989. "Tabu search." *ORSA Journal on Computing*, 1(3): 190–206.

Empirical Evaluation of Local Search Methods for Adapting Planning Policies in a Stochastic Environment

Barbara Engelhardt and Steve Chien

Jet Propulsion Laboratory
California Institute of Technology
4800 Oak Grove Drive
Pasadena, CA 91109
{engelhar, chien}@aig.jpl.nasa.gov

Abstract. Optimization of expected values in a stochastic domain is common in real world applications. However, it is often difficult to solve such optimization problems without significant knowledge about the surface defined by the stochastic function. In this paper we examine the use of local search techniques to solve stochastic optimization. In particular, we analyze assumptions of smoothness upon which these approaches often rely. We examine these assumptions in the context of optimizing search heuristics for a planner/scheduler on two problem domains. We compare three search algorithms to improve the heuristic sets and show that the two chosen local search algorithms perform well. We present empirical data that suggests this is due to smoothness properties of the search space for the search algorithms.

1 Introduction

In many optimization applications, the optimization problem is difficult because of high dimensionality (e.g., large search space) and complex optimization spaces (e.g., non-convex). The problem is made more difficult because the cost of determining the utility of a solution can be expensive (e.g., high computational cost, limited data). The solution cost is exacerbated in stochastic domains where numerous samples are often required to accurately estimate the expected value (which is usually the optimization target) based on a probabilistic decision criterion.

For many large-scale problems, local search and iterative improvement algorithms have been effective in finding good solutions. In particular, many gradient following approaches have been successfully applied to difficult real-world optimization problems [2]. However, these approaches rely on properties of the search space: that the surface has some notion of smoothness to enable a step function, or neighborhood function, to search the space for a local maximum; and that a local maximum is likely to produce an adequate solution. Furthermore, since the particular optimization approach often defines the search operators, it also defines the neighborhood of the strategy search space. Consequently, some optimization approach would result in a search space with smoothness properties while other generation approaches would not.

A. Nareyek (Ed.): Local Search for Planning and Scheduling, LNAI 2148, pp. 108-119, 2001.
© Springer-Verlag Berlin Heidelberg 2001

We examine the effectiveness of local search techniques for difficult optimization problems. We apply local search to learn heuristics to guide search for a planner/scheduler that solves problems from a fixed but unknown problem distribution. We study the effectiveness of local search for optimizing planner strategies, where a strategy encodes the decision policy for the planner at each choice point in the search. In particular, we examine several issues of general interest.

1. We show that two different local search stochastic optimization methods find strategies that significantly outperform both the human expert derived strategy and a non-local search strategy.
2. We show that the smoothness property holds for both local search algorithms (despite their searching two quite different spaces).
3. Surprisingly, examining the learning trials showed that the learning runs had to modify the initial strategies considerably before showing significant improvement. This either meant that the learning algorithms were making poor initial steps, or that the learned strategies lay within a valley. We present empirical results that show that the latter hypothesis is true.

Because our approach is modular to allow arbitrary candidate generation algorithms, we are able to examine the problem for different search strategies. In particular, we examine a local beam-search candidate generation approach and an evolutionary computation approach.

The remainder of this paper is organized as follows. First, we describe the general approach to stochastic optimization. Second, we describe how the planning application is an instance of stochastic optimization. As part of this, we describe the specifics of the control strategy encoding. Third, we describe the empirical results, focusing on the hypotheses outlined above. Finally, we describe related and future work in this area.

1.1 Stochastic Optimization

We define a stochastic optimization problem as optimizing the expected value of a distribution. With a deterministic planner/scheduler and a random set of problems, there is sampling error in the estimation of expected utilities for a set of control strategies, hence the problem is stochastic. A non-deterministic planner/scheduler and a static set of problems will also produce utility distributions with error, hence this problem is also stochastic. In our framework, we have a non-deterministic planner/scheduler *and* a random set of problems, so there is error in estimating each strategy utility distribution, so the problem is stochastic.

We now describe our general iterative framework for optimization of expected value in stochastic domains. First, hypotheses are generated by a local search, then these hypotheses are evaluated by testing them in the application domain and scoring the result (see Figure 1). This testing occurs under the direction of a statistical evaluation component (described below). When the best one or several hypotheses are known with the desired confidence, the process is repeated (e.g., new hypotheses are generated). This entire cycle is repeated until some termination condition is met (e.g., number of cycles, quiescence).

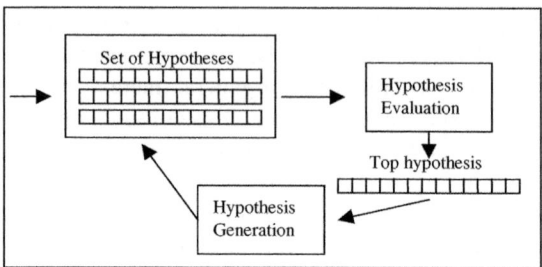

Fig. 1. Optimization cycle - given a set of hypotheses, an evaluation ranks these hypotheses, and a search generates a next generation based on the rank of the previous generation and a candidate generation approach

To evaluate the set of candidate hypothesis steps, we use statistical methods that minimize resources used to satisfy a decision criterion [3][1]. While the algorithm can use an arbitrary decision criterion, in this paper we focus on the use of the Probably Approximately Correct (PAC) requirement, to determine when the utility of one hypothesis is superior to another based on pair-wise comparisons. With the PAC decision requirement, an algorithm must make decisions with a given confidence (expressed as the probability that its selection is correct is greater than δ) to select the appropriate hypothesis (expressed that its expected utility must be within ε of the true best hypothesis) as expressed in Equation (1).

$$\Pr\left[\bigwedge_{i=1}^{n}(\hat{U}(h_i) - \hat{U}(h_{sel})) > \varepsilon\right] \tag{1}$$

Because any specific decision either satisfies or does not satisfy the requirement that the selected hypothesis is within ε of the true best hypothesis, the PAC requirement specifies that over a large number of decisions that the accuracy rate must meet δ. For a pair of distributions, it is relatively straightforward to calculate the probability that one has a higher expected utility than the other. However, selection of a single hypothesis from a set of n hypotheses requires summation of a number of pair-wise comparisons. To minimize resource usage, the algorithm allocates error to each pair-wise comparison based on the estimated cost of samples for those hypotheses, and allocates a greater error to costly comparisons. Thus, the overall error criterion is met using the fewest resources possible by minimizing Equation (2) after each sample where c is the cost of the best hypothesis and the cost of the ith hypothesis, and n is the number of samples allocated to the comparison.

$$\sum_{i=1}^{k-1} c_{sel,i} n_{sel,i} \tag{2}$$

---

[1] In this paper we focus on the candidate hypothesis generation strategies and the outer loop. The statistical evaluation phase of the learning process is described in further detail in [3,4].

The sufficient number of samples (n) can be generated, given a normal distribution of sample utility, by estimating the difference in expected utility and variance of each hypothesis. In general, we cannot solve this problem optimally since the estimates for parameters required to compute optimal solutions will include sampling error. For more information regarding these techniques, see [3].

2 Learning Planner Heuristics as Stochastic Optimization

We investigate stochastic optimization in the context of learning control strategies for the ASPEN planner [5]. ASPEN uses heuristics to facilitate the iterative search for a feasible plan. During each search step, a planner confronts a series of decisions such as which schedule conflict to repair or the action to take to repair it. The planner resolves these choices by stochastically applying the heuristics, based on weights for each choice point heuristic, during iterative repair [17]. Thus the weights define the control strategy of the planner, which impacts the expected utility of the resulting plans.

Specifically, in our setup, a strategy hypothesis is a vector with a weight for each heuristic function and a weight of 0 for a heuristic not in use. The utility of a hypothesis can be determined by running the planner using the control strategy hypothesis on a certain problem instance and scoring the resulting plan. A problem generator for each domain provides a stochastic set of problem instances to enhance the robustness of the expected solution for the entire planning domain.

In our ASPEN setup, there are twelve choice points in the repair search space. Higher level choice points include choosing the conflict to resolve and choosing the resolution method, such as preferring open constraints before violated constraints, or preferring to add activities over moving them. Once a resolution method is selected, further choice points influence applications of the choice point such as where to place a newly created activity and how to instantiate its parameters. For each choice point, there are many heuristics that might be used. The hypothesis vector is the list of relative weight that is given to each heuristic for that choice point. Since the planner is stochastic, the choice of heuristics that are used at each step is randomized, so multiple runs even for the same problem instance may yield a range of solutions (plans) and hence a distribution of utilities.

The search space for each of our domains, given the encoding of the hypotheses, is large. The sum of each choice point's heuristic values must sum to 100 (so each weight can have 101 possible values), and utilities may depend on the correct heuristic values for multiple choice points. So the number of elements in the search space is:

$$\prod_{i=1}^{cp} \binom{101}{h_i - 1},$$
(3)

where h_i is the number of heuristics for choice point i. The two domains we are using have approximately $2.3*10^{30}$ different possible hypotheses. Because there are a limited number of repair iterations (in these experiments, 200 at most), there are a limited number of stochastic decisions to be made, so it is unclear how much of an impact small differences in the weights will make. If we define "small difference" as 10 percentage points for each hypothesis, the space already drops to $4.7*10^{14}$ (substituting 11 for 101 in the above equation) although further sensitivity testing will be done to verify this claim.

2.1 Domains

The repair heuristics were developed for individual domain search requirements from ASPEN applications [5]. There are also domain-specific heuristics, which reference particular features of a domain in order to affect the search. For each domain, the human expert strategy hypotheses were derived independently from (and prior to) our study by manual experimentation and domain analysis.

We examine two different spacecraft domains, which satisfy the normality assumption of the evaluation method. The first domain, Earth Orbiter-1 (EO-1), is an earth imaging satellite. The domain consists of managing spacecraft operations constraints (power, thermal, pointing, buffers, telecommunications, etc.) and science goals (imaging targets and calibrating instruments with observation parameters). Each problem instance is used to create a two-day operations plan: a typical weather and instrument pattern, observation goals (between 3 and 16), and a number of satellite passes (between 50 and 175). EO-1 plans prefer more calibrations and observations, earlier start times for the observations, fewer solar array and aperture manipulations, lower maximum value over the entire schedule horizon for the solar array usage, and higher levels of propellant.

The Comet Lander domain models landed operations of a spacecraft designed to land on a comet and return a sample to earth. Resources include power, battery, communications, RAM, communications relay in-view, drill, and ovens. Science includes mining and analyzing a sample from the comet, and imaging. The problem generator includes between 1 and 11 mining activities and between 1 and 24 imaging activities at random start times. The scoring functions for the Comet Lander domain includes preferences for more imaging activities, more mining activities, more battery charge over the entire horizon, fewer drill movements, and fewer uplink activities.

2.2 Search Methods

The two local search types used were a local beam search method and an evolutionary computation method. The local beam search [13] defines a vector's neighborhood as changing the subset of the vector associated with a choice point by less than a certain step size. As opposed to propagating only highest-ranking vector, the search propagates a beam b of vectors, where b is greater or equal to 1. Samples for each individual candidate hypothesis are generated and scored using the planner, and ranking is done by pair-wise comparisons of these sample utilities for each candidate hypothesis in a generation. For each generation, the beam search takes the top ranking b hypotheses, creates b/g candidate neighbor hypotheses for each of them, and ranks the g candidate hypotheses to create the subsequent generation.

The evolutionary algorithm [7] uses three general operators (crossover, mutation, and reproduction) to generate the next set of hypotheses. Parents are chosen based on their relative ranking, where the higher-scoring hypotheses are more likely to be parents. The crossover operator was not aware of subsets of the hypothesis vector related to each choice point, so it could choose to split within one of those subsets. For all operators, the results are normalized to 100% before evaluation. Samples for each individual candidate hypothesis are generated and scored using the planner, and ranking is done by pair-wise comparisons of these sample utilities for each candidate hypothesis in a generation. For each hypothesis in a generation, the algorithm either reproduces one parent or crosses two parents based on their ranking in the previous generation, and mutates the resulting candidate hypothesis.

Random sampling is another (non-local) method of search. Vectors are generated at random and deep sampling is performed on these vectors for a planning domain. The results show a distribution of random hypothesis points and expected utility for these random points in the strategy space.

Although the local search algorithms are greedy given a correct ranking, due to sampling error the ranking algorithm can produce only an approximation of the correct ranking. Furthermore, as the overall utility of the candidate hypotheses continues to improve, ranking is more difficult because the hypotheses have higher variances relative to the differences in the mean (this is a phenomenon well understood related to the Least Favorable Configuration (LFC) in statistical ranking). Consequently, the highest overall expected utility hypothesis might not occur in the final iteration, and the optimization algorithm does not know the true utilities of the strategies sampled, since it only has estimates. To address this problem, each of our algorithms (beam-search and evolutionary) select the highest estimated utility strategy from all seen during that run (e.g., potentially not the last strategy). When we report that strategy's utility, we report a true utility based on a deep sample of many more samples. Since each run takes several CPU days, we are continuing to perform more optimization runs to provide more detailed results.

3   Empirical Results

One simple question is whether the local optimization techniques improve on the human expert strategies. It is important to keep in mind that the expert strategies were created for a single instance of a problem domain, but in this paper we are evaluating the expert strategies (and the generated strategies) over a set of instances in the problem domain. In both the EO-1 domain and the Comet Lander domain, we compare expected utilities of the handcrafted expert strategy and the best and average strategies found by random sampling (Figure 1). For local beam search and local genetic search we report on the top strategy in the final set of strategies (recall that the beam has several strategies retained and the genetic search has the population) as well as the mean utility of the strategies in the final set. While the learned strategies outperformed the expert strategies, surprisingly the expert strategy in the EO-1 domain was worse than a randomly generated strategy.

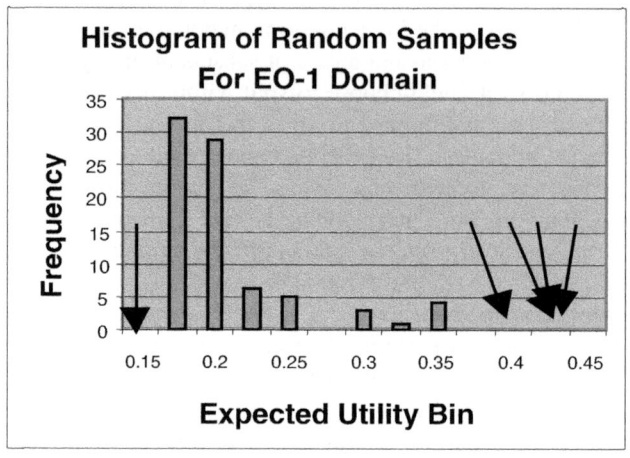

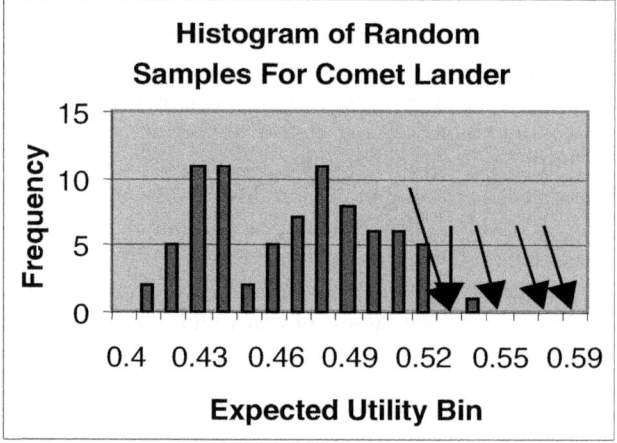

Fig. 2. Histogram summaries

The results show that the local search optimization was able to find strategies that significantly improved on the expert strategies. We plot histograms (Figure 2) for randomly selected strategies in the Comet Lander and EO-1 domains (where the arrows on the histograms indicate key values: expert and learned strategies). These show that the local search optimization techniques found very good strategies overall in the space, among the best possible strategies.

The traces of the two local search techniques operating on each of the domains are shown in the graphs below (e.g., deep sample utility versus iteration). The shapes of these graphs (showing little early improvement) led us to believe that the expert strategies are located in an area of local minima, or a valley, of the search space. In order to test this conjecture, we generated random walks in the strategy spaces. The size of the domain gives us a high probability that a random walk will not cycle. The results show that areas around the starting point perform poorly, and random, undirected steps starting at the expert hypotheses produce little improvement. This data (Figure 3, Figure 5) confirms that the expert strategies lay in a valley but that sufficient gradient information existed to allow the learning to escape the valley. One potential explanation could be that the variance of the problems from a single domain requires a large amount of flexibility in the planner heuristics (e.g. stochasticity), whereas the expert designed the set of heuristics such that it would choose a single non-random strategy for each choice point every time, since the evaluation was on a single instance of a domain.

Table 1. Summary utility results

	Expert	Random Sample		Local Beam Search		Genetic Search	
Domain		High	Mean	High	Mean	High	Mean
Comet Lander	0.538	0.531	0.461	0.584	0.549	0.593	0.569
EO-1	0.161	0.341	0.196	0.446	0.426	0.444	0.382

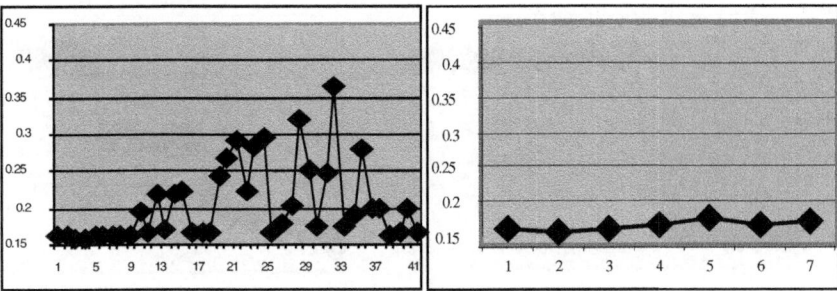

Fig. 3. Two random walks for the EO-1 domain. The first column is beam search, the second is genetic algorithms

How did the local search techniques find their way out of the valley? Local search algorithms are effective on these domains if the search spaces are smooth with respect to the candidate hypothesis generation functions. Smoothness in a discrete domain can be determined by measuring the difference in expected utility between adjacent points with respect to a search step definition. If this difference is small compared to the difference in expected utility between two randomly selected points in the search space, this shows the relative smoothness of the two domains for the search algorithms. For random search, adjacent points are any two vectors in the strategy space. The mean difference in expected utility is measured between two adjacent points, where the initial point is a randomly generated hypothesis, and the adjacent point is one step (as defined by the candidate hypothesis generation function) from that point.

Table 2 shows the adjacency information for the three different methods, which can be considered a measurement of their smoothness. The mean difference between adjacent points shows that two adjacent points from a random sample have four to five times larger difference in utility from adjacent points from the search steps. If the difference in utility is much closer for adjacent points than for random points, a step using search method with this property is likely to remain close to the previous step in terms of utility, so local search methods have a greater chance of being effective.

Table 2. Mean and standard deviations for adjacent points for the three different search methods

Domain	Random Search		Local Beam Search		Genetic Search	
	Mean	Std Dev	Mean	Std Dev	Mean	Std Dev
Comet Lander	0.0435	0.0293	0.0086	0.0066	0.0134	0.0093
EO-1	0.0442	0.0466	0.0114	0.0331	0.0145	0.0244

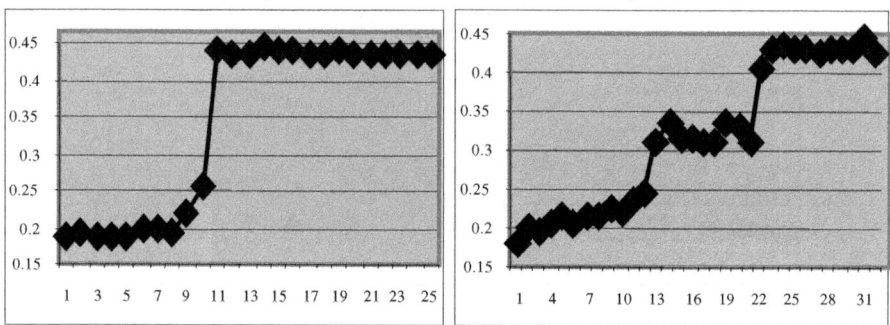

Fig. 4. Two searches for the EO-1 domain. The first column is beam search, the second is genetic algorithms

Although smoothness helps the local search technique step around the search space effectively, using gradient methods is another gamble. We can guarantee by the smoothness analysis that a step will most likely be within some ε of the previous point, but that does not guarantee that improvement using the gradient of that step will allow us to predict the improvement for the next step along that gradient. The data suggests that using gradient methods is effective in finding a path out of the valley, so we believe that some of this gradient information must be preserved in these domains (Figure 4 and Figure 6).

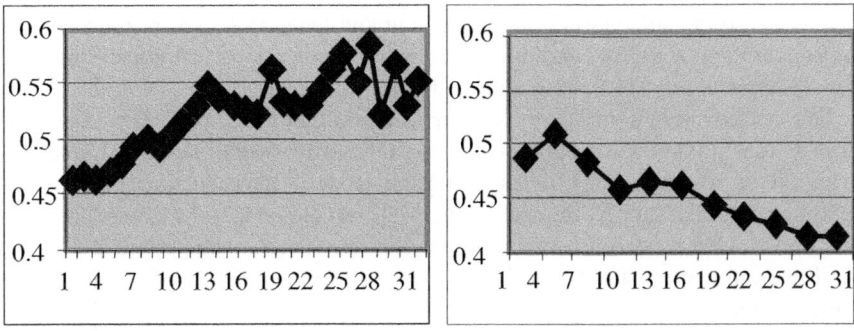

Fig. 5. Two random walks for the Comet Lander domain. The first column is beam search, the second is genetic algorithms

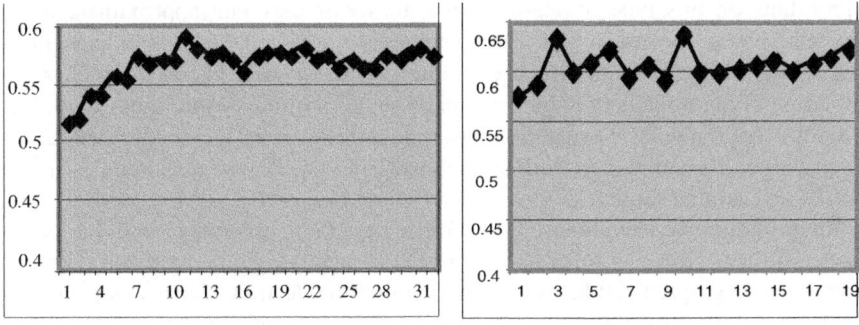

Fig. 6. Two searches for the Comet Lander domain. The first column is beam search, the second is genetic algorithms

4   Related Work, Future Work, and Conclusions

There is significant related work on efficient search techniques. The Q2 algorithm optimizes the expected output of a noisy continuous function, but does not have guarantees on the result [12]. Response Surface Methods [2] have been applied to optimization problems in continuous domains, but require modification for discrete domains (as in our planning heuristics domain). Evaluating control strategies is a growing area of interest. Horvitz [9] described a method for evaluating algorithms based on a cost versus quality tradeoff. Russell, Subramanian, and Parr [14] used dynamic programming to rationally select among a set of control strategies by estimating utility, including cost. MULTI-TAC [11] considers all k-wise combinations of heuristics for solving a CSP in its evaluation, which also avoids problems with local maxima, but at a large expense to the search.

Previous articles describing work in adaptive solving described general methods, which have been developed for transforming a standard problem solver into an adaptive one. Gratch & Chien [7a] illustrated the application of adaptive problem solving to real world scheduling problems and showed how adaptive problem solving can be cast as a resource allocation problem. Zhang and Dietterich used reinforcement learning to learn applicability condition for scheduling operators, using a sliding time window of applicability for those operators [16].

Our optimization approach is similar to learning a naïve bayesian model using an expectation Maximization approach [1,8]. One difference is that our model attempts to minimize resource usage by updating the model after each sample, as opposed to sampling in bulk, simply because of the high sample cost and the low cost to update the model. Our approach is commonly called the frequentist approach, as opposed to the Bayesian approach, since it does not rely on uninformative priors, and instead on initial samples.

We have three major areas for future work. Currently, we are adding simulated annealing cooling functions to determine how to adjust search rates, and if it is effective in this type of search. We would like to test the effectiveness of meta-level learning algorithms on this type of search space, to see if they can approximate the search space and help to evaluate the search methods. Another interesting enhancement is to use a portfolio approach, which combines heuristics and chooses which set to use based on domain features judged statically or at run time. Additional work has been proposed for hypothesis evaluation based on a different set of stopping criteria, which can be resource bounded (specifically considering time as the resource), as in previous works on a similar topic [6].

In this paper we have focused on selecting the planner strategy with the highest expected utility. However other aspects of the strategy might be relevant. For example, consistent (e.g., predictable) performance might be desired. In this case probabilistic decision criteria incorporating undesirability of a high utility variance strategy would need to be used. In particular, the PAC requirement does not incorporate any preference or disliking for high variance strategies.

This paper has presented an approach to optimization of expected values in a stochastic domain is common in real world applications. Specifically, we presented an approach based on local search of the optimization space. We presented empirical results from an application to learning planner heuristics in which learned strategies significantly outperformed human expert derived strategies. And we also presented empirical evidence that these local search techniques performed well because smoothness properties held in these applications.

Acknowledgements. This work was performed at the Jet Propulsion Laboratory, California Institute of Technology, under contract with the National Aeronautics and Space Administration.

References

1. Bishop, C. 1995. Neural Networks for Pattern Recognition. Clarendon Press, Oxford.
2. Box, G.E.P., Draper, N. R. 1987. *Empirical Model-Building and Response Surfaces*. Wiley.
3. Chien, S., Gratch, J., Burl, M. 1995. "On the Efficient Allocation of Resources for Hypothesis Evaluation: A Statistical Approach." In *Proceedings of the IEEE Transactions on Pattern Analysis and Machine Intelligence* 17(7), p. 652-665.

4. Chien, S., Stechert, A., Mutz, D. 1999. "Efficient Heuristic Hypothesis Ranking." *Journal of Artificial Intelligence Research* Vol 10: 375-397.
5. Chien, S., Rabideau, G., Knight, R., Sherwood, R., Engelhardt, B., Mutz, D., Estlin, T., Smith, B., Fisher, F., Barrett, T., Stebbins, G., Tran, D. 2000. "ASPEN – Automating Space Mission Operations using Automated Planning and Scheduling." *SpaceOps 2000*, Toulouse, France.
6. Fink, E. 1998. "How to Solve it Automatically: Selection among Problem-Solving Methods." In *Proceedings of the Fifth International Conf.AI Planning Systems*, 128-136.
7. Goldberg, D. 1989. *Genetic Algorithms: In Search, Optimization and Machine Learning.* Reading, Massachusetts: Addison-Wesley.
7a. Gratch, J.M. and Chien, S. "Adaptive Problem-solving for Large Scale Scheduling Problems: A Case Study," Journal of Artificial Intelligence Research Vol. 4 (1996), pp. 365-396.
8. Heckerman, David. 1996. A Tutorial on Learning with Bayesian Networks. MSR-TR-95-06, Microsoft Corporation.
9. Horvitz, E. 1988. "Reasoning under Varying and Uncertain Resource Constraints." In *Proceedings of the Seventh National Conference on Artificial Intelligence*, 111-116.
10. Lin, S., and Kernighan, B. 1973. "An Effective Heuristic for the Traveling Salesman Problem," *Operations Research Vol. 21.*
11. Minton, S. 1996. "Automatically Configuring Constraint Satisfaction Programs: A Case Study." In *Constraints* 1:1(7-43).
12. Moore, A., Schneider, T., Boyan, J., Lee, M. S. 1998. "Q2: Memory-based Active Learning for Optimizing Noisy Continuous Functions." *Proc. ICML 1998.*
13. Russell, S., Norvig, P. 1995. *Artificial Intelligence: A Modern Approach.* Upper Saddle River, NJ: Prentice Hall.
14. Russell, S., Subramanian, D., Parr, R. 1993. "Provably Bounded Optimal Agents." *In Proceedings of the Thirteenth International Joint Conference on Artificial Intelligence.*
15. Zhang, W., Deitterich, T.G., (1996). "High-Performance Job-Shop Scheduling With a Time-Delay TD(λ) Network *Proc. NIPS* 8, 1024-1030.
16. Zweben, M., Daun, B., Davis, E., and Deale, M. 1994. "Scheduling & Rescheduling with Iterative Repair." In *Intelligent Scheduling.* Morgan Kaufmann. 241-256.

The GRT Planner: New Results

Ioannis Refanidis and Ioannis Vlahavas

Department of Informatics, Aristotle University of Thessaloniki
54006, Thessaloniki, Greece
{yrefanid, vlahavas}@csd.auth.gr

Abstract. This paper presents recent extensions to the GRT planner, a domain-independent heuristic state-space planner for STRIPS worlds. The planner computes off-line, in a pre-processing phase, estimates for the distances between each problem's fact and the goals. These estimates are utilized during a forward search phase, in order to obtain values for the distances between the intermediate states and the goals.

The paper focuses on several problems that arise from the backward heuristic computation and presents ways to cope with them. Moreover, two methods, which concern automatic domain enrichment and automatic irrelevant objects elimination, are presented. Finally, the planner has been equipped with a hill-climbing strategy and a closed list of visited states for pruning purposes. Performance results show that GRT exhibits significant improvement over its AIPS-00 competition version.

1   Introduction

In the last years, some very efficient heuristic planners for STRIPS [5] worlds, like UNPOP [7], ASP [4], HSP [3], HSPr [2], GRT [11] and FF [8], have been presented. These planners work in the space of the states, utilizing domain independent heuristic functions to guide their search. The functions provide quite accurate values for the distances between the intermediate states and either the initial state or the goals. These estimates are computed either on-line, during the search process, or off-line, in a pre-processing phase. Recently, heuristic planning became very popular among the members of the planning community, due to its success in the planning competitions (AIPS-98 and AIPS-00).

The above planners differ mainly in the direction in which they construct their heuristic functions and in the direction in which they traverse the space of the states. These two issues affect significantly the entire problem solving process, i.e. the time needed to solve the problems and the quality of the resulting plans.

The most frequently adopted approach is to construct the heuristic function forward and search for solving a planning problem also forward. This approach has been adopted by the ASP/HSP and FF planners. According to this, in order to estimate the distance between an intermediate state and the goals (while performing forward search), a fast forward search in a relaxed problem, i.e. a problem where actions do not have delete effects, is performed and the number of actions needed to achieve the goals is considered to be this distance. The problem with this approach is that the computation of the distance may take long time, since it has to be repeated for

A. Nareyek (Ed.): Local Search for Planning and Scheduling, LNAI 2148, pp. 120-138, 2001.
© Springer-Verlag Berlin Heidelberg 2001

each state. The advantage is that the estimates obtained in this way can be, under specific conditions, more informative than other approaches.

An alternative approach is to construct the heuristic and search for a plan in opposite directions. This approach has been adopted by the UNPOP, HSPr and GRT planners. Actually, HSPr constructs the heuristic forward and searches backwards, while UNPOP and GRT construct the heuristic backwards and search forward. The advantage of these approaches is that the heuristic is constructed once, in a pre-processing phase, while it is exploited in the search phase for fast distance assignment to the states (actually, UNPOP reconstructs the heuristic for each state, for better estimates). The disadvantage of this approach is that, in some cases, the heuristic has local optimal states (this also happens with the first approach, but more rarely, due to the reconstruction of the entire heuristic function).

Concerning the last approach, the difficult part is always the backward direction. The reason why this direction is problematic is because in most of the problems the goals do not constitute a complete state description, so it is problematic to apply actions to them. HSPr has problems in the search phase with invalid states that arise when regressing the goals. On the other hand, GRT has problems with the heuristic construction, because the incompleteness of the goals makes initially impossible the application of actions to them.

In order to be self-contained, this paper starts with a brief presentation of the GRT planner. Next, we discuss and compare several ways to enhance the goals, in order to make them a complete or a super-complete state description. Then, two methods for accelerating the planning process are presented. The first one concerns the enrichment of the predicate set of a domain, while the second one concerns the elimination of irrelevant to the planning problem objects from the problem description. Finally, we present the adaptation of a hill-climbing strategy to the GRT planner, for quick action selection and application. Comparative performance results show that GRT planner is now significantly more efficient than its earlier version that took part in the AIPS-00 planning competition, and, in some cases, it outperforms FF, the planner that has been awarded in the competition.

2   The Greedy Regression Tables Heuristic

In STRIPS, each ground action a is represented by three sets of facts: the precondition list $Pre(a)$, the add list $Add(a)$ and the delete list $Del(a)$, where $Del(a) \subseteq Pre(a)$. A state s is defined as a collection of ground facts and an action a determines a mapping from a state s to a new state $s'=res(s,a)$. In the formalization used henceforth, the set of constants is assumed to be finite and no function symbols are used, so the set of ground actions is finite. An action a can be applied to a state s, if $Pre(a) \subseteq s$. Then, the resulting state s' is defined as:

$$S' = res(s,a) = s - Del(a) + Add(a) \tag{1}$$

At the beginning of the problem solving process, in a pre-processing phase, GRT estimates once the distances between each problem's facts and the goals. During the search phase, these estimates are used to further estimate the distance between each intermediate state and the goals. Estimating the distance between two states either backwards or forward, often results in different values, since no heuristic is precise. However, in case where the two states are complete, there is no reason why any of the two directions should be preferred.

The estimation of the distances between the problem's facts and the goals is performed by repeatedly applying actions to the goals. Initially the goal facts are considered as achieved and are assigned zero distances. Then, actions are applied to the achieved facts, in order to achieve new facts. Henceforth, when it is noted that a fact has been achieved, this means that it has been assigned a finite distance from the goals. In this section, we consider that the goals are a complete state, so there is no problem to apply actions to them. In the next section we treat the problem of incomplete goal states.

In order to apply actions to the goals, the original problem's actions have to be inverted. Suppose an action a and two states s and s', such that a is applicable in s and $s' = res(s,a)$. The *inverted action* a' of a is an action, such that $s = res(s', a')$. The inverted action is defined from the original action as it follows:

$$Pre(a')=Add(a) + Pre(a) - Del(a)$$
$$Del(a')=Add(a) \qquad\qquad\qquad (2)$$
$$Add(a')=Del(a)$$

GRT applies an inverted action a', if all of its preconditions $Pre(a')$ have been achieved. The newly achieved add effects $Add(a')$ of the inverted action are assigned distances that are a function of the distances of $Pre(a')$. The way in which GRT applies actions is stricter than a usual regression. In the usual regression, an action is applied if a least one of its add effects is included within the facts of a state (here the notion of state refers to all the achieved facts). However, this approach may lead to achieving invalid facts and consequently invalid states. On the other hand, GRT applies an inverted action, if all of its preconditions are within the set of the currently achieved fact.

GRT continues applying inverted actions, until all the problem facts have been achieved. The estimated distances are stored in a table, the rows of which are indexed by the facts of the problem. We call this table the *Greedy Regression Table* (which the acronym GRT comes from), since its data are obtained through greedy regression of the goals.

The GRT heuristic considers the interactions that occur while trying to achieve several facts simultaneously. To each ground fact p of a problem it is assigned not only an estimate of its distance from the goals, but also a list of other ground facts $\{r_1, r_2, ...\}$, which are potentially co-achieved while trying to achieve p. These are facts that are added by some of the actions that achieve p and are not deleted by any subsequent action. This list is called the list of *related facts* of p and is denoted as $rel(p)$. The estimate of the distance between a fact p and the goals is denoted as $dist(p)$.

The estimation of the distances from the goals and the construction of the lists of related facts for all the facts of a problem are illustrated through the steps of Algorithm 1 that follows:

ALGORITHM 1: The main GRT algorithm

Description:	This algorithm estimates the distances from the goals and the lists of related facts for all the ground facts of a problem.
Input:	The action schemas and the predicate definitions of a domain and the objects of a problem.
Output:	The distance estimate $dist(p)$ and the related facts $rel(p)$ for each ground fact p of a problem.

1. Construct the set Inv of all the inverted ground actions of a problem.
2. Let F be the set of all the ground facts of the problem. For each $f \in F$ set $dist(f) = +\infty$.
3. Let $Goals$ be the goal state of the problem. For each $f \in Goals$ set $dist(f) = 0$ and $rel(f) = \varnothing$.
4. While $(Inv \neq \varnothing \land \exists f \in F : dist(f) = +\infty)$ do:
   a) Select an action a from Inv, such that $\forall q_i \in Pre(a)$, $dist(q_i) < +\infty$.
   b) Let $Cost = $ AGGREGATE $(q_1, q_2, \ldots)$, the estimate for the number of steps needed to achieve simultaneously the preconditions q_i of a.
   c) Let $Add'(a)$ be the subset of the ground facts of $Add(a)$ that have not already been achieved:
   $$Add'(a) = \{ p_i \in Add(a): dist(p_i) = +\infty \}$$
   For every $p_i \in Add'(a)$, set $dist(p_i) = Cost + 1$.
   d) For each fact $p_i \in Add'(a)$ compute the list of related facts $rel(p_i)$ as the union of $Pre(a)$, the lists of related facts of a's precondition elements, the other elements in $Add'(a)$, excluding the facts in $Del(a)$. More formally:
   $$rel(p_i) = (\cup_i q_i) \cup (\cup_i rel(q_i)) \cup Add'(a) - Del(a)$$
   e) Remove from Inv all the actions, whose facts in their add-list have all been achieved.

Let us explain the steps of the algorithm. At each iteration an inverted action is selected, the preconditions of which have all been achieved individually (step 4a). Step 4b computes the cost of achieving them simultaneously, step 4c adds 1 to this cost and assigns it to those add-effects of the action that have not already been achieved. Step 4d constructs the list of related facts for the newly achieved facts and finally step 4e removes from the set of the inverted actions, the actions that have not any more facts to achieve. This is repeated until all the ground facts of the problem have been achieved or the set of the inverted actions becomes empty (which is an indication that there are facts with infinite distances from the goals).

The number of iterations performed by Algorithm 1 is bounded both by the number of the ground facts of the problem (since, at each iteration, at least one fact is achieved) and by the number of ground actions of the problem (since, at each iteration, at least one inverted action is removed from Inv). The selection strategy for the inverted actions, which is used in step 4a, affects the results. A good strategy is to select the action with the minimum cost of its preconditions.

Next we present function AGGREGATE. The function takes as input a set of facts $\{q_1, q_2, \ldots, q_n\}$, together with their costs $dist(q_i)$ and their lists of related facts $rel(q_i)$, and estimates the cost for achieving them simultaneously. This function is used both in the pre-processing phase, in order to compute the application cost of the inverted actions, and in the search phase, in order to estimate the distance between each intermediate state and the goals, i.e. the cost of achieving simultaneously the facts of the intermediate state.

Function AGGREGATE

Description: This function estimates the cost for achieving a set of facts simultaneously, by applying inverted actions to the goals.

Input: A set of facts $\{q_1, q_2, ..., q_n\}$, their distances $dist(q_i)$ and their lists of related facts $rel(q_i)$.

Output: An estimate of the cost of achieving the facts $\{q_1, q_2, ..., q_n\}$ simultaneously.

1. Let $M_1 = \{ q_1, q_2, ..., q_n \}$. Let $Cost = 0$.

2. While $(M_1 \neq \emptyset)$ do:

    a) Let M_2 be the set of facts $q_i \in M_1$ that are not included in any list of related facts of another fact q_j of M_1, without q_j being also included in their list of related facts. More formally:

$$M_2 = \{ q_i: q_i \in M_1, \forall q_j \in M_1, q_i \in rel(q_j) \Rightarrow q_j \in rel(q_i) \}$$

    b) Let M_3 be the set of those facts of M_1 that are not included in M_2, but are included in at least one of the lists of related facts of the elements of M_2.

$$M_3 = \{ q_i: q_i \in M_1 - M_2, \exists q_j \in M_2, q_i \in rel(q_j) \}$$

    c) Sum the distances of the facts of M_2. For equivalence classes of facts where each one is included in the list of related facts of the others, consider their common distance once. Add the result to $Cost$.

    d) Let $M_1 = M_1 - M_2 - M_3$.

3. Return $Cost$

In order to explain how function AGGREGATE works, we consider the following example. Suppose that we have three (inverted) actions, a_1, a_2 and a_3, with the following definitions:

a_1: $Prec(a_1)=Goals, Add(a_1)=\{q_1\}, Del(a_1)=\{ \}$
a_2: $Prec(a_2)=\{q_1\}, Add(a_2)=\{q_2\}, Del(a_2)=\{ \}$
a_3: $Prec(a_3)=\{q_1, q_2\}, Add(a_3)=\{q_3\}, Del(a_3)=\{q_1\}$

where $Goals$ stands for all the goal facts and $\{q_1, q_2, q_3\} \cap Goals = \emptyset$. Applying first a_1 to the $Goals$, we find that $dist(q_1)=1$ and $rel(q_1) = \emptyset$. Applying then a_2 we find that $dist(q_2) = 2$ and $rel(q_2) = \{q_1\}$. Finally, applying a_3 we find $dist(q_3) = 3$ and $rel(q_3) = \{q_2\}$.

We will use now function AGGREGATE in order to compute the cost of achieving facts q_1, q_2 and q_3 simultaneously. M_1 initially consists of three facts, $M_1=\{q_1, q_2, q_3\}$. In step 2a of the first iteration, M_2 is set equal to $\{q_3\}$, since q_3 is not included in any list of related facts. The other two facts are not included in M_2, since for example q_2 is included in $rel(q_3)$, while q_3 is not included in $rel(q_2)$. Similarly, q_1 is included in $rel(q_2)$, while q_2 is not included in $rel(q_1)$.

In step 2b, M_3 is set equal to $\{q_2\}$, since q_2 is the only fact that is included in $rel(q_3)$. In step 2c, $dist(q_3)$ is added to *Cost*, so *Cost* becomes 3. Finally, in step 2d, M_1 is set to M_1-M_2-M_3, so M_1 becomes equal to $\{q_1\}$.

The second iteration starts with $M_1 = \{q_1\}$, so consequently $M_2 = \{q_i\}$ (step 2a) and $M_3 = \emptyset$ (step 2b), *Cost* becomes $Cost + dist(q_1) = 4$ (step 2c) and M_1 becomes empty (step 2d). This is the last iteration and value 4, which is the actual cost for co-achieving the three facts, is returned.

Note that, at each iteration, M_2 is partitioned in groups, such that the facts of each group contain one another in their lists of related facts. The facts in each group have all been achieved by the same inverted action and have the same distance. Step 2c sums the costs of these groups.

The number of iterations that function AGGREGATE performs is bounded by the initial size of M_1, but usually, only a single iteration needs to be performed.

2.1   An Example

We illustrate the GRT phases with the *blocks world* problem of Figure 1. Part of the Greedy Regression Table for this problem is shown in Table 1.

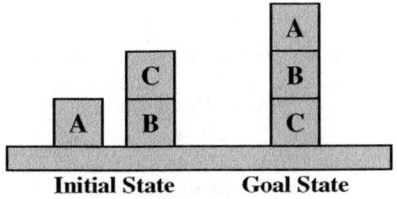

Fig. 1. A 3-blocks problem

Let us compute the distance between the initial state and the goals. The initial state consists of the following facts:

(*on A table*) (*clear A*) (*on B table*) (*on C B*) (*clear C*)

All the above facts are related, with the fact (*on C B*) being the last achieved. So the distance of these facts is the distance of the last achieved, i.e. 3, which in this case is also the actual distance.

Table 1. Part of the Greedy Regression Table for the 3-blocks problem

Fact	Distance from goals	Related facts
(on C table)	0	-
(on B C)	0	-
(on A B)	0	-
(clear A)	0	-
(on A table)	1	(clear B)
(clear B)	1	(on A table)
(on B table)	2	(on A table) (clear A) (clear B) (clear C)
(clear C)	2	(on A table) (clear A) (clear B) (on B table)
(on C B)	3	(on A table) (clear A) (on B table) (clear C)
...	...	...

The above approach is followed to estimate the distances between each intermediate state, which arises during the forward search phase, and the goals. GRT always selects to expand the state with the smallest estimated distance.

2.2 The Utility of the Related Facts

The notion of related facts is unique in GRT, since other similar planners, i.e. ASP and HSP do not use it. Actually, these planners consider that the cost of achieving a set of facts simultaneously is equal to the sum of the costs of achieving these facts individually. However, this approach leads usually to overestimations.

With the introduction of the related facts we try to capture sequences of facts, where each one is a pre-requisite to achieve the next one, without the former being deleted. Performance results have shown to us that plans produced with the use of the concept of related facts are better, both in solution time and in solution length, than in the case of not using them. However, there are cases where exceptions to this general rule arise. In Table 2 we present solution times and lengths for solving some *logistics* problems taken by the AIPS-00 competition, with and without exploiting related facts[1].

Table 2. Solution time/length with and without using related facts (time in msecs).

Problem	With related facts	Without related facts
LOGISTICS-10-0	270/47	320/47
LOGISTICS-10-1	220/43	280/43
LOGISTICS-15-0	660/86	1760/101
LOGISTICS-15-1	550/72	1810/87
LOGISTICS-20-0	1700/120	20820/122
LOGISTICS-20-1	1540/108	33010/113

3 Detecting and Completing Incomplete Goal States

The main problem in the backward construction of the heuristic is that the goals in most of the problems do not constitute a complete state description. For example, in the goals of the *logistics* problems it is not usually determined where the trucks and the planes are. If the goal state is incomplete, it is impossible to apply the inverted actions to it and therefore to construct the heuristic. The solution adopted in GRT is to enhance the incomplete goal states with the missing facts that are not in contradiction with the existing ones. For example, since the goal state of the 'ogistics.a' [13] problem does not determine at which airports the two planes are, it is supposed that each one of the two planes could be at any one of the tree airports. So, the ground facts:

---

[1] All the measurements in this paper have been taken in a machine having an AMD K6-II+ 475MHz processor and 64MB memory, under the MS-Windows 98 operating system and the MS-Visual C++ 6.0 compiler. The last version of the GRT planner is available at the URL www.csd.auth.gr/~lpis/GRT/main.html

(*at plane1 pgh_air*), (*at plane1 bos_air*), (*at plane1 la_air*),
(*at plane2 pgh_air*), (*at plane2 bos_air*), (*at plane2 la_air*).

could be added to the new goal state, referenced henceforth as the *enhanced goal state*.

Two problems arise when trying to convert the goals to a complete state. The first is how to detect the candidate facts that can be used to enhance the goals. The second problem, which is the harder, is to select some of the candidate facts, which will be used in order to enhance the goals. In the next two sub-sections we treat these two problems.

3.1 Detecting the Missing Goal Facts

Concerning the problem of detecting which facts can be used for enhancing the goal state, there are two automatic approaches. The first consists of a forward GRAPHPLAN-like [1] pre-preprocessing phase, where all the binary mutual exclusion relations among the facts of the problem are computed. After this computation, all the non-goal facts that are not mutual exclusive with any goal fact are considered as candidates for being included in the goals. This approach is presented in detail in [12]. Its advantage is that no extra information is needed, apart from the usual STRIPS domain and problem description. Moreover, mutual exclusion relations, which are not easily recognized by a human expert, can be detected. Finally this approach can be also seen as a coarse-grained reachability analysis for the various facts of the problem. The disadvantages of this approach are that it is time consuming and that it does not detect mutual exclusion relations of order higher than two.

The second approach is to use domain specific knowledge in the form of axioms. For example, an axiom can state that a truck or a plane is always located at some place. So, if the goals do not determine where a truck is, it can be deduced a set of candidate final positions of the truck, using this axiom. The advantage of this approach is that the time needed to deduce the candidate facts is negligible, compared to the time needed for the rest of the planning process. Moreover, more complicated relations than simple binary mutual exclusion ones can be encoded. The disadvantage is that extra work is required in the domain encoding. However, there are several approaches for automatic discovery of such domain axioms, with the DISCOPLAN [6] system being the most known.

GRT planner uses the first approach to detect the missing goal facts. So, an overhead in total solution time is imposed by the extra pre-processing work. The burden of this work to the total problem solving time varies from less than 10% in domains like *blocks world*, to more than 20% in domains like *logistics*. The significance of this contribution depends on the difficulty of the domain, i.e. how much time is consumed by the search phase. *Logistics* problems are easier than *blocks world* problem, so in these problems the overhead due to the extra pre-processing work is more sensible. In the future, we intend to adopt an automatic detection of domain axioms method, using analysis of the action schemas and the initial state, in a DISCOPLAN like way, in order to avoid this burden in the total problem solving time.

Table 3 presents the total time needed to solve some of the AIPS-00 planning competition problems and the amount of time spent in the mutual exclusion relations detection. Note than the measurements have been taken with the best search strategies for each domain, i.e. hill-climbing for the *logistics* problems and best-first search for the *blocks world* and *freecell* problems (see Section 6 for the supported search strategies).

Table 3. Time needed to detect the missing goal facts in several problems and domains (time in msecs).

Problem	Total time	Detect missing goal facts time	Ratio
probLOGISTICS-15-1	660	150	22,7%
probLOGISTICS-30-1	9610	1480	15,4%
probBLOCKS8-0	380	50	13,1%
probBLOCKS9-0	1760	50	2,9%
probBLOCKS10-0	830	110	13,2%
probFREECELL-5-1	13020	5310	40,8%
probFREECELL-6-1	84040	6350	7,6%

3.2 Completing the Goals

After the detection of the missing goal facts, some of them have to be selected in order to complete the goals. For example, in the case of problem 'logistics.a', where the goals do not determine where *plane1* is located, a selection has to be made among the candidate facts:

$$(at\ plane1\ pgh\_air), (at\ plane1\ bos\_air), (at\ plane1\ la\_air)$$

GRT supports three different methods, each one of them having different advantages in terms of solution time and quality. These are the following:

1. Select all the facts.
2. Favor initial state facts.
3. Favor the most promising facts, according to the goals.

The first method considers all the candidate facts as goal facts and assigns to them zero distances from the goals. It is obvious that the enhanced goal state that is obtained in this way is not a valid state, since it may contain new facts that are mutual exclusive to each other. The advantage of this approach is that the heuristic computation is very fast, since many facts are achieved at the beginning of the heuristic construction phase and many actions can immediately be applied. The disadvantage of this approach is that the obtained heuristic is less informative, since there is little differentiation in the obtained estimates. The search strategy shares elements from both the best-first and the breadth-first one, it consumes more time but it produces in general better plans than the next two methods.

The second method selects to add to the goals those candidate facts that are also included in the initial state. For example, if *plane1* was initially in *pgh_air*, then the fact (*at plane1 pgh_air*) is added to the enhanced goal state, while the facts (*at plane1 bos_air*), (*at plane1 la_air*), which are mutually exclusive with the selected fact, are rejected. The advantage of this method, with respect to the previous one, is that it results in greater differentiation among the obtained estimates and, consequently, in a faster search phase. On the other hand, favoring the initial state facts is a risk, because if they are not actually included within the goals, the search process may get disoriented, leading to longer plans. However, if there are facts (actually objects' properties) that are not needed for solving a problem, this approach lets them unaffected, thus facilitating the search process.

The third method tries to combine the advantages of the first two. In contrast to them, where the enhancement of the goals is performed in a single step, before the construction of the heuristic, this method adds facts to the goals in a progressive manner, in parallel with the heuristic construction. Actually, new facts are added to the enhanced goal state only in the case where no new applicable inverted actions exist. In this case one or more of the candidate facts are assigned zero distances, until a new inverted action satisfies its preconditions.

The way in which the candidate facts are selected is not random, but there are some rules that favor specific facts among others. Specifically, facts that can be combined with already achieved facts in order to make an inverted action applicable are preferred. The following four rules are applied in decreased preference:

• First are selected the facts that can be combined with the original goal facts.
• Next are selected the facts that can be combined with other already achieved facts.
• Next are selected the facts that are included in the initial state.
• Finally, the remaining candidate facts are selected in a random way.

This method can take also advantage of the information of the actions that firstly achieved the facts during the forward GRAPHPLAN-like phase, where the mutual exclusion relations were computed. So, the first two of the above rules give priority to the candidate facts that render applicable the actions which achieved forwards the already achieved facts. Note that in all cases, when a candidate fact is selected, other candidate facts that are mutually exclusive with the selected one are removed from the set of the candidate facts.

The third method is the default method of the GRT planner and results in general to the best solution times, producing also in many cases equal or better plans than the other two methods. However, especially in terms of plan quality, there are many exceptions that depend on the specific problem. In general, it is not difficult to construct problems for any one of the methods presented above, where the method performs better than the others.

Table 4 presents performance results in terms of solution time and length for the above three methods and for several *logistics* problems used for the AIPS-00 competition. As we can see, best solution times are obtained by the second and the third method, while the first one is always the slowest. Concerning solution length, we can see that the three methods have uniform behavior, each one of them being better in some problems. For all the problems of Table 4 a best-first strategy has been used.

In other domains, like the *blocks world*, the *freecell* or the *miconic-10* ones from the AIPS-00 competition, or the *gripper* and the *movie* domains from the AIPS-98 competition, the goals are complete or near-complete state descriptions, so in these domains the method selected for goals completion plays no significant role. In other domains, like the *mystery* one (AIPS-98), the only acceptable strategy for goal completion is the first one, since many of the candidate facts are really unreachable.

Table 4. Solution time/length for the three goals completion methods (time in msecs)

Problem	All facts	Initial facts favored	Most promising facts
ProbLOGISTICS-13-0	1980/**82**	**1380**/85	1650/87
ProbLOGISTICS-13-1	1480/67	**1100/66**	1150/69
ProbLOGISTICS-14-0	1760/65	**990/61**	1090/66
ProbLOGISTICS-14-1	1590/**76**	**1210**/79	1260/77
ProbLOGISTICS-15-0	1980/**82**	1810/83	**1370**/83
ProbLOGISTICS-15-1	4880/**72**	1270/75	**1260/72**
ProbLOGISTICS-20-0	33890/118	3760/120	**3630/114**
ProbLOGISTICS-21-0	49540/113	**3400**/114	4510/**110**

Note finally that the enhanced goal state is used only in the pre-processing phase, when the heuristic function is constructed. During the search phase the planner tries to reach a state where only the original goals are included. So, completeness is never lost. However, the way in which the heuristic function has been constructed favors implicitly specific search paths and thus it may disorient the search process.

4   Domain Enrichment

In this section, we present an approach adopted by the GRT planner, in order to cope with poor domain descriptions. With the word 'poor' we refer to domains where negative facts are implicitly present in the initial state and in the preconditions of the actions. We faced this problem twice, with the *movie* domain of the AIPS-98 competition and with the *miconic-10* domain of the AIPS-00 competition.

Let us consider first the *miconic-10* domain. In this domain, there is an elevator, several floors and several passengers. Each passenger is in an initial floor and has to be moved to her/his destination floor. This domain is described by four action schemas, the (*board ?floor ?passenger*) and (*depart ?floor ?passenger*) for boarding on and leaving the elevator and the (*up ?floor1 ?floor2*) and (*down ?floor1 ?floor2*) for moving the elevator.

Action (*board ?floor ?passenger*) is defined by the following formula:

```
(:action board
  :parameters (?f ?p)
  :precondition (and (floor ?f) (passenger ?p)
    (lift-at ?f) (origin ?p ?f))
  :effect (boarded ?p))
```

As we can see, the only dynamic predicate in the definition of action *board* is predicate *boarded*, which denotes that the passenger has boarded the elevator, whereas all the preconditions of this actions are static. The problem with this kind of action definition is twofold. First, the action can be applied many times for the same passenger in the same plan, i.e. a passenger may board the elevator although she/he is already boarded! Second, and more specific to GRT, there is no mention that a passenger has not boarder. Actually, the initial state contains implicitly a fact for each passenger denoting that he/she is not yet boarded. However, this is not stated explicitly, while GRT takes into account only explicitly stated facts when computing distances between states and the goals. So, the initial states of the *miconic-10* problems contain only static facts, which are not removed in the successor states, and thus they are assigned zero distances from the goals!

What is needed is the definition of a new predicate, for example *not_boarded*. Facts of this predicate should be added to the initial state, denoting that each passenger has not boarded. Moreover, action *board* should be changed accordingly, in order to include these facts in its preconditions and remove them after its application.

GRT performs the above transformations at run-time. The identification of the above situation is performed in a way similar to the identification of the incomplete goals. In this case, GRT looks for dynamic facts of a problem that are not mutual exclusive with any initial state fact. If this happens, the negative facts of the identified ones are defined at run-time and added to the initial state. Furthermore, the actions that add the positive facts are transformed, so that the negative facts appear in their preconditions and delete effects.

In the *miconic-10* domain this situation arises with the *board* and *depart* actions and the *boarded* and *served* predicates. GRT defines the *not_boarded* and *not_served* predicates at run-time, it enhances the initial state with facts determining that each passenger is neither boarded nor served yet and actions *board* and *depart* are transformed accordingly. For example, action *board* is transformed to the following definition:

```
(:action board
  :parameters (?f ?p)
  :precondition (and (floor ?f) (passenger ?p)
   (lift-at ?f) (origin ?p ?f) (not_boarded ?p))
  :effect (and (not (not_boarded ?p))(boarded ?p))
```

A similar situation arises in the *movie* domain of the AIPS-98 competition. In this domain, the goal is to have enough snacks in order to watch a movie. There are several actions of the form:

```
(:action get-chips
          :parameters (?x)
          :precondition (and (chips ?x))
          :effect (and (have-chips)))
```

This action has as precondition the static fact (*chips ?x*) and produces the dynamic fact (*have-chips*). The action can be applied several times, however once only is enough to achieve the goal of having chips. The problem in this domain is that the initial state implicitly declares that we do not have chips (and dips and pops etc), but no specific dynamic facts make this clear. So, in case where no domain enrichment process takes place, GRT would assign to the initial state a zero distance from the goals.

In this case GRT detects that there are problem facts, like (*have-chips*), (*have-dips*) etc, that are not mutual exclusive with any fact of the initial state, it defines their negative facts, it adds them to the initial state and it transforms accordingly the actions.

In both of the domains mentioned in this section, the GRT planner without the domain enrichment technique cannot solve the problems but some of the easiest ones. On the other hand, using this technique it can solve all the problems quite easily.

5 Eliminating Irrelevant Objects

In many domains there are objects that are irrelevant to any solution. The most typical examples can be found in the transportation domains, like *logistics*, *mystery* or *miconic-10*, where some packages are initially in their destinations, or for which no specific destination is determined. If we remove these objects, together with all the facts and actions where these objects appear, from the problem description, we gain in terms of efficiency due to the reduced size of the problem, whereas we do not lose completeness.

We developed in GRT a method that detects and removes irrelevant objects, which do not necessarily appear in any solution. The method concerns pure STRIPS domains without negation in the preconditions of the actions or in the goal formula, however it can be easily extended to cover these cases. The objects are identified before the pre-processing phase, using the following two rules:

An object is irrelevant to any solution of a specific planning problem if:
- It does not appear in any fact of the goals, except for the case where the same fact is also included in the initial state, and
- there is no action having this object in its preconditions, except for the case where the object appears also in all of the action's effects.

The above conditions are very strict, but they ensure that any detected object is certainly irrelevant, so they preserve the completeness of the problem solving process.

Proposition. Any object satisfying the above rules can safely be removed from the problem description, without sacrificing the completeness of the problem solving process.

Proof: Suppose that we have identified an object *obj* for which the above two rules hold. We will show that this object is not necessary for achieving any other goal fact. Suppose that there is a goal fact $g \in Goals$ not containing *obj*. Suppose also that there is an action that achieves *g*, which has a precondition containing *obj*. In this case, the second rule is violated, since there is an action having *obj* in its preconditions, without *obj* appearing in all of its effects. So, fact *g* can be achieved only by actions that do not have preconditions depending on *obj*. Thus, if we regress *g* using an action, the produced subgoals would not contain *obj*. However, in the same way we can reject actions that have *obj* in their preconditions and achieve other non-goal facts that do not contain *obj*, i.e. the subgoals produced after *g*'s regression. So, *obj* is not necessary for achieving any goal or subgoal of the problem. On the other hand, there is no fact containing *obj* within *Goals*, that has to be achieved, or if there is one, it is already present in the initial state, according to the first rule. So, *obj* can safely be removed from the problem description. ∎

With the above rules we can safely and fast identify irrelevant objects in several domains. We present the efficiency of the above technique with some *logistics* problems from the AIPS-00 competition, where we added colors. More specifically we added a dynamic predicate (*painted ?object ?color*) denoting the color of a package, a static predicate (*color ?color*) for the definition of the available colors and an action (*paint ?object ?old_color ?new_color*) for changing the color of a package. We defined four colors and we assigned a color to each package of the initial state. The goal state does not determine the colors of the packages. So, in this case the colors are irrelevant objects and can safely be removed, along with all the facts and actions that include colors. We ran GRT with and without the irrelevant objects elimination technique on these problems and the results are shown in the Table 5.

Table 5. Solution time/length with and without eliminating irrelevant objects.

Problem	With irrelevant object elimination	Without irrelevant object elimination
probLOGISTICS-13-0	3840/82	5870/82
probLOGISTICS-13-1	2690/67	4070/67
probLOGISTICS-14-0	3070/65	3790/65
probLOGISTICS-14-1	2860/76	3290/76
probLOGISTICS-15-0	3570/82	4120/82
probLOGISTICS-15-1	3240/72	3780/72

The application of the two rules for the elimination of the irrelevant objects can be done in a progressive manner. Suppose that in the example with the colored packages there are also brushes that have to be used in order to perform the paint operation. So, there are two new actions, (*get ?brush*) and (*leave ?brush*) and the predicate (*have ?brush*), which is an add effect of the *get* action. The predicate (*have ?brush*) is also a precondition in the enhanced action (*paint ?package ?color ?brush*). In this case, the brushes are also irrelevant and should be removed from the description of the problem. However, since action *paint* needs brushes and has effects that do not contain brushes (i.e. (*painted ?package ?color*)), brushes are not removed, due to the violation of the second rule. However, after removing all the color objects, the *paint* actions are also removed, so brushes do not violate the second rule for the remaining actions and can be removed safely.

The disadvantage of the proposed is approach for the elimination of irrelevant objects is that it does not remove objects that can eventually appear in a plan, but for which there are other better (i.e. shorter) plans that do not need these objects. For example, in the *logistics* domain, suppose that we have three cities, *city1*, *city2* and *city3*, each one of them having a truck to move packages between its various locations, there is a plane that can fly between the airports of the cities and a package that has to be moved from a location of *city1* to a location of *city2*. In this case, *city3*, together with its locations and its truck are not necessarily needed to solve the planning problem, since the package can be moved directly from *city1* to *city2*, without going via *city3*. However, it is not easy to detect the irrelevance of *city3*. Actually, there are plans that move the package from *city1* to *city2* through *city3*. If we decide to remove *city3* and its objects from the domain representation, we take the risk of sacrificing completeness, since probably the problem may become unsolvable. Deciding safely that *city3* and its objects can be removed, without sacrificing completeness, can be as hard as solving the original problem itself.

Another approach for ignoring irrelevant facts and actions (not objects) has been proposed some years ago in [10]. In this work the elimination is based on heuristics that approximate a plan by backchaining from the goals and ignoring any conflict. However, this approach, although achieving more elimination, is not solution preserving and sacrifices completeness. Furthermore, this approach may be more time consuming, since it demands the construction of an initial approximate plan.

6 Hill-Climbing Strategy

We equipped the GRT planner with a second, optional search strategy, the well-known hill-climbing. Other heuristic state-space planners, like HSP and FF, take advantage of this strategy, which sacrifices completeness in order to gain efficiency.

In our implementation, at each intermediate state the hill-climbing strategy looks for an action to be added at the end of the current plan, such that the resulting state has a smaller estimated distance from the goals than the previous one. GRT does not try all the applicable actions to the state, but only those actions that achieved the facts of the state during the backward heuristic construction phase. This is similar to the 'helpful' actions of the FF planner, although in the case of GRT it is not sure that all the tried actions are applicable. In case where no better state is found, the rest of the applicable actions to the current state are tried. If again no better state is found, a bounded breadth-first search is performed, with a maximum depth defined by the user (the default value is 6).

As soon as an improving state is found, the actions are added to the end of the current plan and the hill-climbing search continues from the new state. In case where the bounded breadth-first search does not find a solution, GRT starts from the initial state with the best-first search algorithm, which is complete.

Table 6 presents performance results for some *logistics* and *miconic-10* problems, using the hill-climbing and the best-first strategies. As for the *logistics* problems the most promising facts goal completion method has been used, whereas for the *miconic-10* problems the all facts one.

Table 6. Solution time/length for the hill-climbing and the best-first strategies for several *logistics* and *miconic-10* problems (time in msecs).

Problem	Hill-climbing	Best-first
LOGISTICS-15-0	660/86	1270/83
LOGISTICS-15-1	550/72	1370/72
LOGISTICS-20-0	1700/137	6210/114
LOGISTICS-20-1	1490/114	3570/110
LOGISTICS-25-0	3960/166	88820/158
LOGISTICS-25-1	3960/177	79800/158
MICONIC-10-0	440/34	490/34
MICONIC-20-0	2970/70	3570/70
MICONIC-30-0	20370/101	27470/101

As we can see, in the *logistics* problems the acceleration of the hill-climbing strategy is significant, especially in the large problems. The cost for this acceleration is the longer plans. In the *miconic-10* problems the acceleration is less significant, while the plans found by the two search strategies are identical.

Results for other domains, like *blocks world* and *freecell*, are not presented, since in these domains the hill-climbing strategy changes quickly to the best-first one, due to the inability to find improving successor states.

7 Performance Results

In this section, we present new performance results of the GRT planner for some of the AIPS-00 planning competition problems. We compare the planner to its previous version that took part in the competition, and to FF, which has been awarded in the domain independent planning track.

The new features of the GRT planner are the following:

- Several ways to complete the goals.
- Irrelevant objects elimination.
- Hill-climbing strategy.
- Closed list of visited states.

All the measurements have been taken in an AMD K2+ 475 MHz computer, having 64 MB memory, under the MS-Visual C++ 6.0 compiler. We compiled the planners with the 'maximize for speed' option /O2 (in the AIPS-00 competition GRT was running without any compiler optimization). The source code of GRT is available at www.csd.auth.gr/~lpis/GRT/main.html, while the source code of FF is available at http://www.informatik.uni-freiburg.de/~hoffmann/ff.html (special thanks to Joerg Hoffman for making his code available).

Table 7 presents comparative results in some *logistics* problems. We ran the new GRT with the all facts goals-completion method, since we observed that with the hill-climbing strategy there is no significant difference in the planning times, while there is a difference in the plan lengths, for the various goals-completion methods. The new GRT planner is in many cases faster than FF, however FF produces in general shorter plans. In this domain, the acceleration of the new GRT to the old one is mainly due to the hill-climbing strategy. Short dashes indicate that no solution is found after 5 minutes.

Table 7. Indicative solution time/length results for some *logistics* problems (time in msecs).

Problem	Original GRT	New GRT	FF
LOGISTICS-10-0	500/46	270/46	270/46
LOGISTICS-10-1	440/43	220/43	220/42
LOGISTICS-20-0	58330/118	1700/120	2410/118
LOGISTICS-20-1	48890/113	1540/108	1640/111
LOGISTICS-30-0	-	10220/198	10930/186
LOGISTICS-30-1	-	9610/208	20550/200
LOGISTICS-40-0	-	60750/252	55200/251
LOGISTICS-40-1	-	63330/250	112050/241

Table 8 presents comparative results in some *blocks world* problems. In these problems the goals are complete state descriptions, so it makes no difference which goals-completion method is used. GRT cannot solve these problems with the hill-climbing strategy and after some trials it turns to the best-first strategy. In the *blocks world* problems the original GRT failed to solve most of the problems, except for some of the easiest. The new GRT takes advantage of the closed list of visited states and performs efficient pruning, thus it solves more problems. However, there remain problems that are unsolvable.

The GRT heuristic is inefficient with this kind of *blocks world* domain definition, i.e. using the actions *stack*, *unstack*, *push* and *pop*. The difficulty comes from the fact that all the movements are performed in two steps and the intermediate state with the block being held by the arm is common for many different movements. So there is very little differentiation in the estimated distances between the facts and the goals. If *move* actions were used instead, GRT would produce much better plans and would solve larger problems.

FF performs well in this domain and this is due to a technique called *Added Goal Deletion*, according to which the goal facts are ordered and achieved in a progressive manner [8, 9]. This technique is especially suited for the *blocks-world* domain and the 4-operators representation. However, this technique does not always succeeds to produce good orderings and this is the reason why FF fails to solve some of the easiest problems, which have been solved by GRT.

Table 8. Indicative solution time/length results for some *blocks world* problems (time in msecs).

Problem	Original GRT	New GRT	FF
BLOCKS-6-0	820/36	110/40	50/20
BLOCKS-6-1	710/18	50/15	50/10
BLOCKS-6-2	1100/32	110/32	60/20
BLOCKS-7-0	1890/22	160/22	110/20
BLOCKS-7-1	89620/56	220/56	110/22
BLOCKS-7-2	243760/52	270/48	280/22
BLOCKS-8-0	> 5 mins	380/38	110/18
BLOCKS-8-1	95450/44	330/44	-
BLOCKS-8-2	19560/50	280/50	50/16
BLOCKS-9-0	-	1760/116	-
BLOCKS-9-1	-	1710/104	220/28
BLOCKS-9-2	560	280/44	110/26
BLOCKS-10-0	-	830/108	110/34
BLOCKS-10-1	-	61950/108	3790/38
BLOCKS-10-2	-	49760/124	270/34
BLOCKS-11-0	-	550/50	110/34
BLOCKS-11-1	-	2530/134	-
BLOCKS-11-2	-	2030/128	170/34
BLOCKS-12-0	-	43610/114	940/44
BLOCKS-12-1	-	47790/68	160/34
BLOCKS-13-0	-	-	220/41
BLOCKS-13-1	-	-	-
BLOCKS-14-0	-	-	220/40
BLOCKS-14-1	-	-	720/42
BLOCKS-15-0	-	-	-
BLOCKS-15-1	-	-	330/52
BLOCKS-16-1	-	-	-
BLOCKS-16-2	-	-	280/52
BLOCKS-17-0	-	-	-

Table 9. Indicative solution time/length results for some *freecell* problems (time in msecs).

Problem	Original GRT	New GRT	FF
FREECELL-2-1	2360/10	2200/9	490/9
FREECELL-3-1	4780/25	4230/16	1210/21
FREECELL-4-1	-	6650/24	3740/26
FREECELL-5-1	-	13020/29	5220/37
FREECELL-6-1	-	84040/38	11320/43
FREECELL-7-1	-	60590/47	8950/48
FREECELL-8-1	-	40480/54	20320/48
FREECELL-9-1	-	37460/64	17080/73
FREECELL-10-1	-	150730/89	103590/87

Table 9 presents some results from the *freecell* domain. Also in this domain, the original GRT failed to solve problems due to the absence of a closed list of states. The new GRT planner succeeds to solve almost all problems (the correction of some bugs concerning the instantiation of the actions in the original GRT has been proved very helpful in this domain). Concerning solution time, FF performs better. However, concerning solution length, most of the times GRT finds better plans.

Finally, Table 10 presents performance results in the *miconic-10* domain. In this domain, the original GRT planner is slower than the new one, mainly due to the different search strategies, however the two versions produce exactly the same plans. FF is faster by a constant factor of about 3.

Table 10. Indicative solution time/length results for some *miconic-10* problems (time in msecs).

Problem	Original GRT	New GRT	FF
MICONIC-15-0	2190/49	1260/49	330/46
MICONIC-20-0	5440/70	2970/70	1020/64
MICONIC-25-0	19060/88	7880/88	2250/89
MICONIC-27-0	28010/95	11140/95	2750/89

8 Summary

This paper presented recent extensions to the GRT planner, a domain independent heuristic state-space planner for STRIPS worlds. In a pre-processing phase GRT computes estimates for the distances between each fact of a problem and the goals. These estimates are used to further estimate the distances between the states of the state-space and the goals, thus guiding the search process in a forward direction.

The paper treated four different issues. First, several methods for detecting and completing incomplete state descriptions have been presented. The reason for this work was the need to complete the usually incomplete goal states, to make it possible to apply actions to them, in order to construct GRT's heuristic. We presented two methods for detecting candidate goal facts, one based on a GRAPHPLAN-like mutual exclusion relations computation phase and one based on the exploitation of domain axioms. Moreover, we presented three methods to select some of the candidate facts, in order to be included in the goals.

Next, an approach for enriching poor domain representations has been presented. This method can be applied successfully in domains, where negative predicates are implicitly present in the initial state and in the preconditions of the actions. We presented how these predicates can be detected and how a domain can automatically be enriched, in order to represent them explicitly.

Next we presented a rigorous method for detecting and eliminating irrelevant objects, together with the facts and the actions where the irrelevant objects appear. The method is based on analyzing the initial state, the goals and the actions and is solution preserving.

Finally, we enhanced the GRT planner with a hill-climbing strategy and with a closed-list of visited states. Performance results for the new GRT planner have shown its superiority against its earlier version, which took part in the AIPS-00 planning competition. Moreover, GRT has been compared to the FF planning system, which has been awarded in the domain independent track of the last competition. The measurements have shown that the two planners are very close to each other, however in many cases FF remains faster.

References

1. A. Blum and M. Furst, Fast Planning Through Planning Graph Analysis, Artificial Intelligence **90** (1997) 281-300.
2. B. Bonet and H. Geffner, Heuristic Planning: New Results, in: Proceedings 5th European Conference on Planning, Durham, UK, LNAI 1809, Springer (1999), pp. 360-372.
3. B. Bonet and H. Geffner, HSP: Heuristic Search Planner, entry at the AIPS-98 Planning competition, Pittsburgh, 1998.
4. B. Bonet, G. Loerincs and H. Geffner, A robust and fast action selection mechanism for planning, 14th Intl. Conf. of the AAAI, Providence, AAAI Press, 1997, pp. 714-719.
5. R.E. Fikes and N.J. Nilsson, STRIPS: A new approach to the application of theorem proving to problem solving, Artificial Intelligence **2** (1971) 189-208.
6. A. Gerevini and L. Schubert, Inferring State Constraints for Domain-Independent Planning, 15th International Conference of the AAAI, Wisconsin: AAAI Press, 1998, pp. 905-912.
7. D. McDermott, Using regression-match graphs to control search in planning, Artificial Intelligence **109** (1-2) (1999) pp. 111-159.
8. J. Hoffmann, A Heuristic for Domain Independent Planning and its Use in an Enforced Hill-climbing Algorithm, Technical Report No. 133, Institut für Informatik, Freiburg, January 2000.
9. J. Koehler, and J. Hoffmann, On Reasonable and Forced Goals Orderings and their Use in an Agenda-Driven Planning Algorithm, Journal of Artificial Intelligence Research, **12**(2000) pp. 339-386.
10. B. Nebel, Y, Dimopoulos and J. Koehler, Ignoring Irrelevant Facts and Operators in Plan Generation, 4th European Conference on Planning, 1997.
11. I. Refanidis and I. Vlahavas, GRT: A Domain Independent Heuristic for STRIPS Worlds based on Greedy Regression Tables, 5th European Conference on Planning, Durham, UK, LNAI 1809, Springer (1999), pp. 347-359.
12. I. Refanidis and I. Vlahavas, On Determining and Completing Incomplete States in STRIPS Domains, IEEE International Conference on Information, Intelligence and Systems, Washington, US, 1999, pp. 289-296.
13. M. Veloso, Learning by Analogical Reasoning in General Problem Solving, Ph.D. diss. Computer Science Dept., Carnegie Mellon Univ. (also available as techincal report: CMU-CS-92-174), 1992.

Incremental Local Search for Planning Problems

Eva Onaindia, Laura Sebastia, and Eliseo Marzal

Dpto. Sistemas Informaticos y Computacion
Universidad Politecnica de Valencia
Camino de Vera s/n, 46022 Valencia, Spain
{onaindia,lstarin,emarzal}@dsic.upv.es

Abstract. We introduce a new approach to planning in STRIPS-like domains based on an incremental local search process. This approach arises as an attempt to combine the advantages of a graph-based analysis and a partial-order planner. The search process is carried out by a four-stage algorithm. The starting point is a graph, which totally or partially encodes the planning problem. The aim of the second phase is to obtain a first set of actions of a solution plan, the third stage guarantees the completeness and optimality of the generated solution and the fourth stage, a partial-order planner, completes the process by finding the missing actions of the final solution plan, if any.

1   Introduction

Graphplan-like or SATPLAN-like planners have shown to outperform classical planners for most of the standard planning domains. However, these two propositional approaches do not exhibit good results for large-sized problems due to the size of the graph they have to deal with. We tested STAN [3] and Blackbox [6] on large problems from the *blocksworld* domain and noticed that none of them were able to solve problems involving more than fifteen blocks.

Our motivation is to develop a new planning approach, which also offers a good performance for large-sized problems. In order to tackle this issue, we introduce a search method, which integrates a technique that incrementally exploits the problem knowledge and a Partial-Order Planner (POP). Our method is executed in four stages:

- The aim of the first stage is to generate a graph containing a set of actions. This graph may include all actions of a solution plan.
- The result of the second stage is a more refined graph which only contains a subset of actions which will necessarily appear in a correct solution.
- The third stage takes the solution obtained from the previous stage and returns a new improved partial plan. The purpose of this stage is to guarantee the completeness and optimality of the generated solution, i.e., to ensure that (a) this partial solution will eventually lead to a final plan, if a solution exists for the given problem, and (b) the plan will be optimal. This is achieved by finding a partial consistency order between the action nodes in the graph.

A. Nareyek (Ed.): Local Search for Planning and Scheduling, LNAI 2148, pp. 139–157, 2001.
© Springer-Verlag Berlin Heidelberg 2001

At this stage unsolvable problems are detected and optimal solutions are found. In some cases, the graph resulting from this phase will comprise all the actions of a final solution plan.
- The fourth stage implements a POP which is aimed at adding the missing actions for the final plan and finding a total ordering relation among all the actions in the plan.

We present here a local search approach to planning which, starting from an initial graph generated from the problem, iteratively improves the partial plan contained in this graph. The second stage obtains a plan which may be a final valid solution or which possibly contains some unsatisfied preconditions and/or inconsistent ordering relations between actions. Then the third and fourth stages repair the flaws in the plan by choosing the "best" choice for each flaw. This is done by applying different criteria based on the graph properties and heuristics to measure the appropriateness of one choice with respect to another.

2   Creating the Problem Graph

The first phase of the algorithm creates a graph inspired in a Graphplan-like expansion. This graph, named *Problem Graph* (PG), may partially or totally encode the planning problem. The PG is a directed, layered graph with two kinds of nodes (literals and actions) and two kinds of edges (precondition-edges and add-edges). The levels alternate action levels containing action nodes and literal levels containing literal nodes.

- An action level A_j consists of all action instantiations a_{jk} which satisfy these two requirements:
  - all the preconditions of a_{jk} are present in the previous literal level L_{j-1} and
  - a_{jk} does not occur in any previous action level
- A literal level L_j is a set of propositions implictly representing the different world states reachable after executing the actions in A_j. More specifically, the set of literals in L_j is defined as $L_{j-1} \cup \mathsf{AddEff}(a)^1$ $\forall a \in A_j$, being a an action instantiation in A_j.

The first level in the PG is the literal level L_0 and it is formed by all the literals in the initial situation. A_1 consists of all action instantiations which are applicable in L_0. L_1 is the set of literals in L_0 plus the add effects of each action in A_1 and so forth. The PG creation terminates when a literal level containing all the literals from the goal situation is reached in the graph or when no new actions can be applied.

It must be noticed that a PG is neither a state-space graph nor a Planning Graph [1]. There are two main differences with respect to a Planning Graph:

---

[1] AddEff, DelEff and Pre stand for the add effects, delete effects and preconditions of an action respectively.

a) Levels in the PG do not stand for time steps but for instantiation steps which can comprise more than one execution step. An action level A_j denotes that all the actions in A_j will be executed at a time step $t \geq j$, and at least one action from A_{j-1} must be executed firstly.

b) Our PG does not take into account mutual exclusion relations between actions, so two actions in a level may interfere with each other, be complementary or independent actions. The PG is created by a forward-chaining process which simply adds the positive effects of actions.

For all the tested domains (see Section 5), except the *hanoi* problem, all the necessary actions of a valid solution already appeared in the PG. This cannot be always guaranteed because, as it was said above, the PG generation terminates when all the literals from the goal situation are present in a literal level, even though additional actions could be applied in this final level. The way of creating the PG ensures that the majority of propositions that would be generated with a systematic search method are obtained once the final literal level is reached.

The advantage of the PG is that its size is much smaller than the Planning Graph and the cost of creating this graph is hardly appreciable even when dealing with large-sized problems.

In the following, we illustrate the process of creating the PG for the *Sussman* anomaly problem in the *blocksworld* domain. Table 1 shows the initial literal level L_0 which consists of one node for each proposition in the initial situation. Action level A_1 contains the two applicable actions in the initial situation, unstack C A and pickup B, and L_1 all the literals at L_0 plus the add effects of the two actions at A_1. The column to the right of A_1 shows the numbers given to the preconditions and add effects of each action (P stands for preconditions and E stands for effects).

The process carries on with action level A_2 (Table 2). At this level, seven action instantiations, different from those in A_1, are found. The literal level L_2 is created by adding the add effects of the actions at A_2 (literals numbered from 10 to 14). This table also shows the final action and literal levels. The two literals

Table 1. Problem graph for the *Sussman* anomaly (1)

L_0		A_1		L_1	
on A table	1	unstack C A	P={4,5,6} E={7,8}	on A table	1
clear B	2	pickup B	P={2,3,6} E={9}	clear B	2
on B table	3			on B table	3
on C A	4			on C A	4
clear C	5			clear C	5
arm-empty	6			arm-empty	6
				holding C	7
				clear A	8
				holding B	9

Table 2. Problem graph for the *Sussman* anomaly (2)

A_2		L_2		A_3		L_3	
pickup A	P={1,6,9} E={10}	on A table	1	unstack C B	P={5,6,12} E={2,7}	on A table	1
putdown B	P={8} E={2,3,6}	clear B	2	unstack B C	P={2,6,13} E={5,8}	clear B	2
putdown C	P={7} E={5,6,11}	on B table	3	unstack B A	P={2,6,14} E={8,9}	on B table	3
stack C B	P={2,7} E={5,6,12}	on C A	4	stack A B	P={2,10} E={6,9,15}	on C A	4
stack C A	P={7,9} E={4,5,6}	clear C	5	stack A C	P={5,10} E={6,9,16}	clear C	5
stack B C	P={5,8} E={2,6,13}	arm-empty	6	putdown A	P={10} E={1,6,9}	arm-empty	6
stack B A	P={8,9} E={2,6,14}	holding C	7			holding C	7
		holding B	8			holding B	8
		clear A	9			clear A	9
		holding A	10			holding A	10
		on C table	11			on C table	11
		on C B	12			on C B	12
		on B C	**13**			**on B C**	**13**
		on B A	14			on B A	14
						on A B	**15**
						on A C	16

from the goal situation, on B C and on A B, are present at literal level L_3 and the PG creation is completed.

3   Description of the Search Method

The search process itself is carried out in two stages. The output from both stages is a graph, each representing an improved plan.

3.1   The Basic Graph

The objective of this phase is to find the appropriate actions to satisfy the goal literals, then the actions to satisfy the subgoals (preconditions) of the former actions and so on. The result is a directed, layered graph with only action nodes named *Basic Graph* (BG). The number of levels in the BG is the number of action levels in the PG plus two additional levels, an initial and a final action level. The former contains one action with effects and no preconditions and the latter contains one action with preconditions and no effects. The effects of the initial action a_0 are the literals in the initial situation and the preconditions of the final action a_n are the goal literals. An edge or causal link $a_i \rightarrow a_j$ denotes that a precondition of a_j is solved by means of an add effect of a_i.

The process starts with the preconditions of a_n and attempts to find a set of actions in the same or any previous action level having these goals as add effects. The preconditions of these actions form a new set of subgoals and this operation is repeated until each literal has been processed.

In order to find a consistent causal link for each subgoal, the search method applies the following property:

Property 1 (literal consistency). *A literal p required by an action a_m ($p \in$ Pre(a_m)) is said to be consistent if these two requirements hold:*

1. *there is a sequence of actions $a_i \rightarrow a_{i+1} \ldots a_{m-1} \rightarrow a_m$ such that $p \in$ AddEff(a_i) and $p \notin$ DelEff(a_j) $\forall j \in [i+1, m-1]$.*
2. *for each action a_k such that $p \in$ DelEff(a_k) there is a sequence $a_k \rightarrow a_{k+1} \ldots a_{m-1} \rightarrow a_m$ with an action a_l, $l \in [k+1, m-1]$, such that $p \in$ AddEff(a_l).*

The first part of the property states that there must exist an action with p as an add effect and no actions deleting p must appear after that action (Figure 1(1)). The second part states that for each action which deletes p there must exist an action ordered after it which produces p (Figure 1(2)).

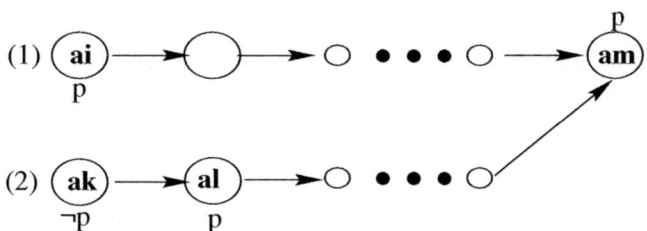

Fig. 1. Literal consistency property. Literals above each node represent the preconditions of actions and literals below the nodes denote add and delete effects.

In order to check literal consistency it is necessary to propagate effects of an action a_i each time a causal link $a_i \rightarrow a_j$ is asserted. The propagated effects of an action a_j are computed by means of the following procedure:

1. PDelEff(a_0) =DelEff(a_0)
   PAddEff(a_0) =AddEff(a_0)
2. Let $p_0, p_1, \ldots, p_n$ be the paths in the graph that have a_j as destination node. Let $A = \{a_{0,j-1}, a_{1,j-1}, \ldots, a_{n,j-1}\}$ be the set of predecessor actions of a_j, each corresponding to a path.

a) $\mathsf{PAddEff}(a_j) = \{x \in \mathsf{PAddEff}(a_i) : a_i \in A/(\exists a_k \in A \wedge x \in \mathsf{PDelEff}(a_k)) \rightarrow a_k < a_i\ ^2\}$

$\mathsf{PDelEff}(a_j) = \{x \in \mathsf{PDelEff}(a_i) : a_i \in A/(\exists a_k \in A \wedge x \in \mathsf{PAddEff}(a_k)) \rightarrow a_k < a_i\}$

b) $\mathsf{PAddEff}(a_j) = \mathsf{PAddEff}(a_j) - \mathsf{DelEff}(a_j) \cup \mathsf{AddEff}(a_j)$

$\mathsf{PDelEff}(a_j) = \mathsf{PDelEff}(a_i) - \mathsf{AddEff}(a_j) \cup \mathsf{DelEff}(a_j)$

The literal consistency property determines that a precondition p of an action a_i is consistent if the propagation of effects from the initial action a_0 to a_i returns that $p \in \mathsf{PAddEff}(a_i)$ and $p \notin \mathsf{PDelEff}(a_i)$ at step 2(a) from the above procedure. Figure 2(a) shows an example of a consistent precondition, p, for the action a_k and Figure 2(b) shows that the precondition p of action a_k is still an inconsistent literal.

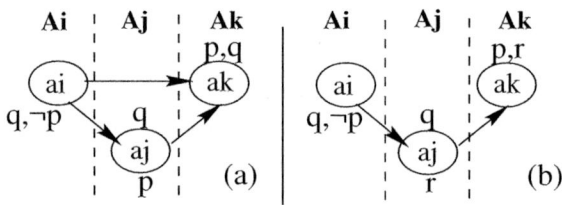

Fig. 2. Example of a consistent and an inconsistent literal

The literal consistency property guarantees that, when a problem is solvable, all the actions in the BG form a part of a correct solution, although all necessary actions may not appear in the BG. However, it is not possible to prove that such a solution exists or that the set of actions leads to an optimal solution. This will be the objective of the next stage.

Basically, the BG represents a plan which contains optimal sequences of actions to achieve each subgoal literal independently. If property 1 does not hold for all action preconditions, then it means that some of the subgoal literals cannot be satisfied and consequently the BG is not created. Let's take the example in Figure 2(b). If p stands for *arm-empty* and the only action having p as an add effect is the initial action a_0 (a_i in the example) and there is not an operator *putdown*, then the BG cannot be created.

The conflicts detected at this stage are only those which appear in a same sequence of actions. Conflict between actions of different sequences will be discovered at the next stage. Finally, it must also be noticed that there might be actions in the PG which belong to a correct solution and are not discovered during the generation of the BG.

---

^{2} $a_k < a_i$ denotes an ordering relation between a_k and a_i such that a_k is executed before a_i

Figure 3 shows the BG for the *Sussman* anomaly problem. Action stack A B is introduced in the BG at A_3 to satisfy the precondition 15 of the final action a_n, and action stack B C is included in the BG at A_2 to satisfy literal 13 of a_n. Notice that these are the only actions in the PG which satisfy literals 13 and 15 respectively.

The algorithm proceeds now with the preconditions of the two actions inserted in the BG. Action stack A B requires literals 2 and 10. Literal 10 is only achieved by action pickup A whereas literal 2 can be produced by several actions in the PG in action level A_2 or by action a_0. The link pickup A $\rightarrow$ stack A B is inserted to satisfy literal 10, and the resolution of literal 2 is postponed until more information is available. Similarly, the action to satisfy the precondition 8 of stack B C is clearly identified whereas there are more than one choice for literal 5 in the PG (several actions in action level A_2 or action a_0).

The introduction of action pickup B to satisfy the precondition 8 of action stack B C causes precondition 2 of action stack A B to become an inconsistent literal, as the propagation of effects deletes this literal. Then it is necessary to find an action between A_1 and A_3 to restore literal 2; since there is already an action at A_2 in the BG which has literal 2 as an add effect (action stack B C), this action is chosen to solve precondition 2 of stack A B.

Following the same process, action unstack C A is introduced to satisfy literal 9 of action pickup A. As literal 5 of stack B C is not removed by any action (the propagation of delete effects from unstack C A does not cause any conflict because the action stack B C is not reached), the algorithm selects a_0 to satisfy this literal.

Finally, literal 6 of pickupA is solved by means of action stack B C; the remaining preconditions of unstack C A and pickup B are all solved with a_0. It is important to notice that the algorithm always selects at first place actions already contained in the BG rather than introducing a new action in the BG from the PG. The criteria the algorithm applies to select an action for a literal when there are several choices are explained in [11] in more detail.

3.2   The Optimal Graph

This stage performs two different tasks taking the BG as input:

- To discover whether the problem is solvable.
- To verify that the plan comprised in the BG leads to an optimal solution plan.

In order to accomplish these two tasks, the search method applies the following property.

Property 2 (partial consistency). *A BG is partially consistent if it is possible to set a total-order relation between each pair of actions in the same action level of the BG.*

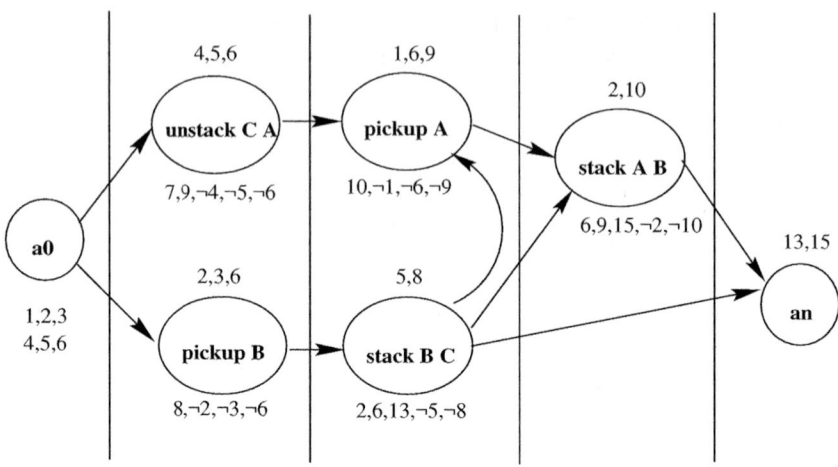

Fig. 3. BG for the sussman anomaly problem

Definition 1 (mutual exclusion). *Two actions a_i, a_j in the same level are mutually exclusive between each other if a_i deletes a precondition of a_j and a_j in turn deletes a precondition of a_i.*

We will show that the application of property 2 guarantees the completeness of the search method (it can find a graph if a solution exists for the given problem and returns "no graph" otherwise) and optimality of the obtained solution (optimality refers to the total number of actions in the final plan). The task of ensuring the total consistency of the graph is carried out by the POP at the fourth stage.

Unsolvable problems. Let a_1 and a_2 be two mutually exclusive actions, $\mathsf{Pre}(a_1) = \{x_1, y\}$, $\mathsf{Eff}(a_1) = \{z_1, \neg y\}$, $\mathsf{Pre}(a_2) = \{x_2, y\}$, $\mathsf{Eff}(a_2) = \{z_2, \neg y\}$.

If the only possible way to satisfy z_1 and z_2 is by means of actions a_1 and a_2 respectively, then we say this is a *precondition conflict*. The name comes from the fact that the literals involved in the conflict are the preconditions of actions (in the example, a_1 needs literal y and deletes y and likewise for a_2). In this case, the literal in conflict has to be achieved again by a new action (from the PG) or an existing action (from the BG). The only additional checking is to discover the correct ordering for the new producer action a_3 ($a_1 \to a_3 \to a_2$ or $a_2 \to a_3 \to a_1$).

Let's assume the precondition conflict cannot be solved with any of the actions in the PG or BG, that is neither Property 1 nor 2 hold after the resolution process. As there may be missing actions in the PG, the algorithm extends the PG one additional level to search for new actions to solve the conflict. This process is repeated until the PG cannot be further extended, that is, until no new action instantiations can be applied. Notice the PG only adds positive effects whose producer actions have not been previously asserted in the graph and therefore it is usually the case that only a few additional levels will be generated. Once the PG cannot be further expanded we can ensure all different action instantations are already included in the PG. Thus when a precondition conflict cannot be solved by any means the problem is unsolvable.

Theorem 1 (unsolvable problem). *If a problem is unsolvable then one of the following choices occurs:*

1. *any of the goal literals never appear in the PG or*
2. *the BG cannot be created or*
3. *there exists an irresoluble precondition conflict in the BG.*

Proof. A problem is unsolvable if there does not exist a sequence of actions which being applied to the initial situation gives rise to the goal state. Let O be the set of operators of a problem, $L = \{p_1, p_2, \dots, p_n\}$ the set of positive literals which are generated by all the instantiations of each $o \in O$ and G the set of goal literals.

(1) If such a sequence of actions does not exist then it might be the case that $\exists p_i \in G$ and $p_i \notin L$, in whose case p_i will never appear in the PG.

(2) If such a sequence of actions does not exist then it might be the case that it is not possible to find an independent sequence of actions for each $g \in G$, in whose case the BG will not be created because property 1 fails.

(3) If such a sequence of actions does not exist then it might be the case that the BG does contain a sequence of actions for each $g \in G$ but it is not feasible to combine all of them to form a global solution. In this case an irresoluble precondition conflict will be found in the BG as property 2 will eventually fail. Notice that establishing an ordering constraint between a pair of actions is equivalent to apply a resolution method such as promotion or demotion, and the procedure to solve mutually exclusive actions is equivalent to the process to restore a deleted literal in a POP. Thus, all resolution methods to solve a conflict are considered.

The above theorem does not state that these three cases are the only situations under which a problem turns out to be unsolvable. However, from our experience with this first prototype of the search method we can conclude that, if one of the above choices holds, then the problem is unsolvable. Let's explain this in an informal way.

If the **PG cannot be created** is because a given goal $g1$ never appears in the PG. Then the problem is unsolvable because:

- either there is no operator having $g1$ as an add effect in the domain definition, in whose case the problem has no solution or
- there exists an operator whose instantiation would generate $g1$ but it is never applicable during the PG creation. It is important to notice that the PG is a relaxed, unrestricted graph where the delete effects of actions are not taken into account. Consequently, if the conditions necessary for the operator to be applicable do not hold in the PG they will not hold in a plan either.

The situations under which the **BG cannot be created** do not all necessarily lead to an unsolvable problem. If the BG cannot be created it means that at least one inconsistent literal is found, i.e. property 1 fails because it is not possible to find a sequence of actions for achieving that literal. There are two different explanations for this situation:

- If p only appears at L_0 (initial literal level) ($p \notin L_h$ /$h \in [1,k]$) then there is no applicable operator having p as an add effect, in whose case p cannot be satisfied and property 1 fails. In this case it is easy to see the problem is unsolvable.
- On the other hand, if ($p \notin L_h$ /$h \in [1,k]$) it might occur there exists an applicable operator but this appears at a level A_l where $l > k$. Since our algorithm only considers action levels equal or lower than the one of the needer action, it would not find such an add effect to achieve the precondition p. In this case, the algorithm would return the problem is unsolvable although there might be a solution for it. This inconvenient can be easily tackled by considering all action levels when creating the BG, that is when solving the action preconditions.

The last case is when an **irresoluble precondition conflict** is found in the BG. Property 2 attempts to find a consistent ordering for each pair of actions a_i, a_j in the same level. If such an ordering exists (for example, $a_j < a_i$), then it means that a_i must be executed after a_j - one time step further than a_j at earliest. Then a_i is moved forward to the next action level and the same operation is repeated again to search for new mutually exclusive actions. This is an indirect way to verify that the sequences of actions that achieve each subgoal separately can be properly combined. Therefore, when an irresoluble precondition conflict is found it means there is a sequence of actions for achieving each subgoal but not a correct ordering for the entire set of actions. Since all the actions in the PG, BG and extended PG (if necessary) are taken into account when solving a conflict, we can conclude that an irresoluble precondition conflict entails an unsolvable problem.

In summary, if one of the three conditions holds then the problem is usually unsolvable. However, in some cases the algorithm could report that no solution exists for a solvable problem. All this indicates the algorithm is able to discover all unsolvable problems and might fail at finding the solution for some solvable problems.

Theorem 1 states that, when dealing with an unsolvable problem, one of the three conditions will hold. So, if the BG is created for a problem which is known to be unsolvable, then the lack of a solution will be discovered at the time of solving precondition conflicts. In order to illustrate this, let's take the following example from the *blocksworld* domain. The initial situation is the same as for the *Sussman* anomaly, that is, on C A, on A table and on B table. The goal state consists of the two contradictory literals: on B C and on C B.

The PG for this unsolvable problem is the same as for the *Sussman* anomaly problem until literal level L_2 (Table 1). The two literals from the goal situation

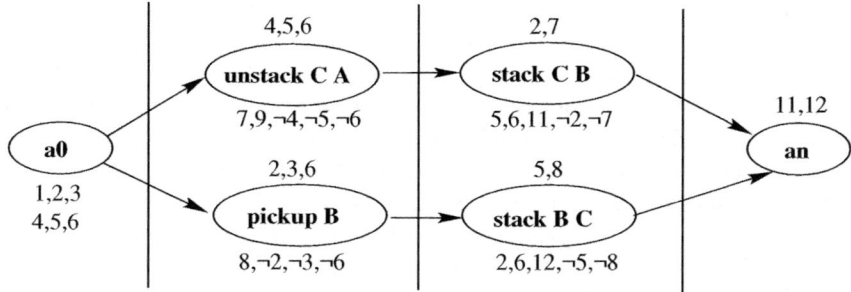

Fig. 4. BG for the unsolvable problem

are found in L_2 and the process for creating the PG terminates. The corresponding BG is shown in Figure 4.

As it can be seen in Figure 4, there exists a sequence of actions for each action precondition. However, we find a conflict between the two actions in A_1 as they both require and delete literal 6 (arm − empty).

When solving the precondition conflict we find six potential actions in the PG in action level A_2 to solve literal 6. Actions stack C B and stack B C cannot be used to restore literal 6 because they would both make property 1 fail (action stack C B would delete literal 2 which is required by pickup B, and action stack B C would delete literal 5 which is required by unstack C A). The other four actions make property 2 fail as all of them require and delete literals 7 or 8 which are also required by the actions in the BG (they are all mutually exclusive with the two actions stack C B and stack B C in the BG in action level $A2$).

The PG is then extended one level further to find new actions to solve the conflict (Table 2). Three of the actions in A_3 have literal 6 as an add effect. Action stack A B cannot be used because it requires and deletes literal 2 and therefore it cannot be ordered between unstack C A and pickup B. The same happens with action stack A C, which needs and deletes literal 5. The remaining choice is to use action putdown A which requires literal 10. As this literal is only generated by action pickup A, which in turn deletes literal 6, property 2 would fail because the same conflict is found again in the graph. The PG can be extended one more level (A_4 and L_4) and no actions for solving literal 6 are found at this level. We can conclude that the precondition conflict is irresoluble and consequently the problem has no solution.

Optimal solutions. During the process of creating the BG and verifying whether the problem is solvable, the selection of an action to solve a literal tends to obtain the minimal set of actions [11].

Property 3 (minimal graph). *Let A be the set of actions in a graph G. G is a minimal graph if for each precondition p of an action $a_k \in A$ there exists only one sequence of actions $a_i \rightarrow a_{i+1} \ldots a_{k-1} \rightarrow a_k$ such that $p \in$ PAddEff(a_{k-1})*

This property is also extended as follows: when there are several actions to solve a literal, the algorithm selects first an action whose preconditions are all already solved with the nodes in the graph instead of selecting an action which in turn needs additional actions to satify its preconditions. The application of property 3 is a first step to obtain an optimal plan. However, some additional checking must be done in order to guarantee the optimality.

Let a_1 and a_2 be two mutually exclusive actions, $\mathsf{Pre}(a_1) = \{x_1, y\}$, $\mathsf{Eff}(a_1) = \{z_1, \neg y\}$, $\mathsf{Pre}(a_2) = \{x_2, y\}$, $\mathsf{Eff}(a_2) = \{z_2, \neg y\}$. If there are other choices to generate $z1$ or/and $z2$ then we say this is an *effect conflict*. This problem usually arises due to a lack of information during the creation of the BG. The algorithm will then replace the producer action of $z1$ (or $z2$) by another action in the BG or PG which has this literal as an add effect. It is important to notice that the first operation carried out by the search method is to verify the effect conflict rather than the precondition conflict. This is because the resolution of a precondition conflict always inserts a new action instead of replacing one of the two mutually exclusive actions.

Proposition 1 (optimal graph). *Let G be a basic graph and A be the set of actions in G. If property 2 holds for the set A then G is an Optimal Graph (OG), that is A is a subset of the actions of an optimal solution.*

Proof. (1) The PG consists of the minimum number of levels as possible. (2) The process of creating the BG always tends to select the minimal set of actions (given two goals g_1 and g_2, if action a_1 achieves g_1 and action a_2 achieves both g_1 and g_2 then only a_2 is selected). (3) Actions existing in the BG are preferred to actions in the PG to solve conflicts caused by mutually exclusive actions. All this means that the algorithm is aimed at obtaining the optimal sequence of actions for each subgoal. (4) If there are no mutually exclusive actions, then it is feasible to succesfully combine the sequences of actions for each subgoal to produce an optimal plan.

Two remarks must be commented about this process:

- An OG leads to an optimal solution plan provided that the POP is known to be admissible.

- The third stage only verifies a partial consistency in the BG, which is very helpful to discover whether the problem is unsolvable and to check the optimality of the generated partial solution. However some other conflicts, as negative threats, can still be present in the OG (because they are not detected at the third stage). Thus the task of the POP will be to solve these conflicts and ensure the total consistency of the final solution plan.

The application of property 2 does not only allow to discover unsolvable problems but also to find and repair some conflicts in the BG. Figure 5 shows the OG for the *Sussman* anomaly problem after applying property 2. In the BG of Figure 3 we can see there are two mutually exclusive actions at A_1: both actions unstack C A and pickup B require and delete literal 6 (arm − empty). On solving this conflict, the algorithm searches for actions in the PG at the next action level A_2, to restore literal 6. There are several choices:

1. action putdown B is mutually exclusive with action stack B C, which is already in the BG
2. stack C B would make literal 2 of action stack A B become inconsistent
3. stack C A is mutually exclusive with action pickup A, which is already in the BG
4. stack B C cannot be used to restore literal 6 because it would delete literal 5 of unstack C A
5. stack B A is mutually exclusive with stack B C
6. putdown C does not cause any conflict

The algorithm selects putdown C to solve the conflict and the result is the OG shown in Figure 5.

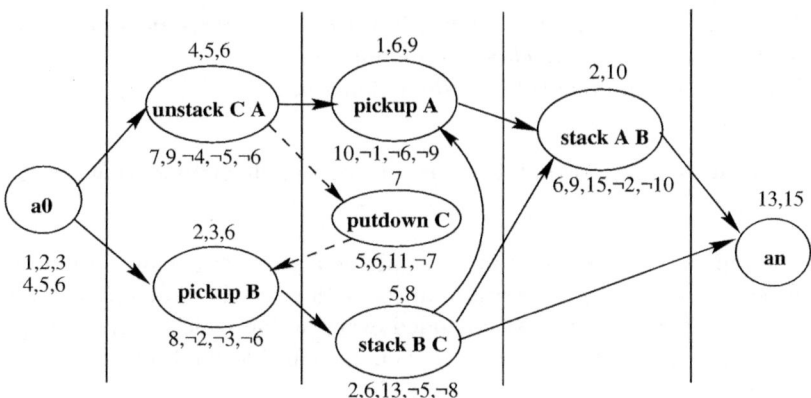

Fig. 5. OG for the *sussman* anomaly problem

4   The Partial-Order Planner

Our partial-order planner [9] is based on the UCPOP planner [8] and therefore completeness is guaranteed when starting from an empty initial plan. However, when the input to the POP is not an empty plan, additional operations are necessary to guarantee the completeness of the planner. In this section we show how to achieve this task. On the other hand, the planner uses an admissible heuristic search [4] which guarantees obtaining an optimal plan.

It must be pointed out that a POP cannot guarantee terminating on unsolvable problems although the complete search space is explored. Since this task has already been performed by the third stage we can ensure that the POP is only executed when it is confirmed a solution exists for the problem. The only remaining task is then to guarantee the POP is able to find such a solution.

4.1   From a Graph to an Input Graph

This section presents the process to generate an input plan from the BG or OG. This plan will be the initial plan for the Partial-Order Causal-Link (POCL) planner.

Let $\mathcal{G}(\mathcal{N}, \mathcal{E})$ be a graph where $\mathcal{N}$ is the set of graph nodes so that $\mathcal{N}$ is a subset of the set of actions that constitute a solution plan for the given problem, and $\mathcal{E}$ is the set of edges that represent causal links between the actions in $\mathcal{N}$.

Definition 2 (Input plan). *An input plan is a tuple* $\Pi(\Lambda, \Sigma, \Theta, \Phi, \Gamma)$, *where:*

– $\Lambda = \mathcal{N}$ *is a set of actions that belong to the input plan.*
– $\Sigma = \mathcal{E}$ *is a set of causal links between the actions in* Λ.
– Θ *is a set of ordering constraints between the actions in* Λ, *resulted from the causal links in* Σ.
– Φ *is a set of non-satisfied preconditions of the actions in* Λ *(agenda) which in the case of an input plan is an empty set.*
– Γ *is a set of conflicts between the actions in* Λ.

There are two important features of our search method that must be enhanced:

– The input plan may not comprise all the actions of a solution plan because either the PG does not contain all the actions of the problem or the process of creating the BG has not included every action which might belong to the solution. The POCL planner is then responsible for finding these missing actions.
– In the process of creating the BG, when finding an action to solve the precondition of another action a_j, only action levels A_t where $t \in [0, j]$ are analyzed. This is because, according to the PG creation, it is more likely to find the correct action in a lower action level than in an upper level. However, since action levels do not stand for execution steps it might be the case that the correct action to solve a precondition of a_j were in an action level A_t where $t > j$. This issue is partially overcome at the OG stage since at this phase all action levels are taken into account.

These two features make the search method focus on a very restricted search space at the expense of introducing non-correct causal links for some literals (that is, the correct producer action to satisfy the literal is not the one denoted in the causal link).

When the POP input is an empty plan, a complete search space is generated and all choices to solve an open precondition or a conflict are considered in the resolution process. However, when the input is not an empty plan, completeness is not guaranteed because this non-empty plan is just the result of one branching line of the search space which would have been generated by a complete search method.

A way to recover completeness in the POP is by means of the White Knight (WK) concept [2]. This technique simply provides a combination of promotion, demotion, separation and precondition establishment methods.

Our WK technique [10] consists in deleting the threatened causal link and insert the precondicion associated to the causal link into the agenda (which it will be satisfied in turn by another existing or new action). This technique can also be used when a threat cannot be solved by promotion or demotion. Let a_i, a_j, a_k be three actions in the input plan such that $a_i \to a_j$, $a_i \to a_k$, $p \in \mathsf{AddEff}(a_i)$, $p \in \mathsf{Pre}(a_j)$, $p \in \mathsf{DelEff}(a_j)$, $p \in \mathsf{Pre}(a_k)$ and $p \in \mathsf{DelEff}(a_k)$. The following process is used to restore the precondition p of a_j or a_k (symmetrical threats):

1. select one of the symmetrical threats $\gamma(a_k, a_i, a_j)$ and solve it by promotion and/or demotion
2. select the other threat $\gamma(a_j, a_i, a_k)$ and use the WK technique
3. repeat this process inverting the order at which the threats have been selected

Proposition 2 (POP completeness). *If the plan obtained from the BG or OG is input to the POP as the initial plan, then the POP will find a solution.*

Proof. It is easy to check that all choices to restore the non-correct causal link are considered in this process. If the correct action for a literal is an action in the input plan, this will be discovered at step 2 of the above process when selecting an existing action. If the correct action does not appear in the PG, and consequently neither in the BG nor in the OG, this will be discovered at step 2 when selecting a new action. Thus, completeness is recovered during the planning process carried out by the POP by using a WK-based technique.

4.2   Main Features of the POP

It is important to distinguish between the two different types of plans that are input to the POP. The WK technique is mainly used when a precondition p of an action a_j is deleted by one action $a_i, i < j$ which belongs to another sequence of actions. However, in other graphs it is possible to solve interactions among actions by just finding a consistent ordering among them. This gives rise to two different types of graphs and, consequently, to two different types of plans. Let $\mathcal{A}$ be the set of actions in a graph and $\mathcal{S}$ the set of actions that constitute a solution plan for a given problem:

- *complete graphs*, when $\mathcal{A} = \mathcal{S}$. In this case, it is possible to set a total-order relation among all the actions in $\mathcal{A}$ without restoring any literal.
- *incomplete basic graphs*, when $\mathcal{A} \subset \mathcal{S}$. In this case, the WK technique must be used to solve the conflicts and new actions will have to be added in the input plan.

It is important to remark the similarity of the process undertaken at the third and fourth stage. In both cases the goal is to restore non-correct causal links in the graph. Unlike the third stage, the POP uses a WK technique when the appropriate actions to solve the conflicts are missing actions that do not appear in the PG; that is the reason why the third stage is not able to detect such conflicts.

The goal of the POP is to find a final plan, which involves two main tasks: detecting and solving conflicts among the actions in the OG and adding the necessary actions to complete the plan. These missing actions belong to any of the following three types:

- actions which do not appear in the PG or
- actions which appear in the PG but they are not discovered during the creation of the BG or
- actions which do appear in the PG and more than one execution of the same action is required (repeated actions).

5   Experimental Results

The experiments shown in table 3^3 correspond to a previous prototype of our method where the third stage is not implemented. Therefore, the algorithm is not obtaining the optimal solution for problems such as monkey test 1, nor detecting unsolvable problems.

Problems were taken from the UCPOP suite and Blackbox software distribution. All tests were run on a Sun Ultra 10 machine and results are given in seconds. We solved each problem ten times with Blackbox v3.6 [6] - using the Graphplan search engine-, STAN [3] and our search method. The results are classified into two groups, those for complete graphs and those for incomplete graphs (Table 3).

In most of the problems where the search method was able to obtain a complete graph, the CPU time was reduced by more than 50% compared to STAN and Blackbox. For example, in the *blocksworld* domain, as the number of blocks increased, this difference was greater. This is specially noteworthy in *TowerLarge* problems: our method solved the *TowerLargeD* problem that neither STAN nor Blackbox were able to solve.

For those problems with an incomplete graph, the search method behaves slightly worse that STAN and Blackbox, although this difference was not as significant as in the previous case. As the average and standard deviation results show (Table 4), our method behaviour was much more stable.

6   Related Work

Recently, local search techniques have been applied to planning problems. In particular, Kautz and Selman developed a local-search based method (Walksat)

---

[3] GT stands for the time used in the graph creation and TT for the total time. We have used a blocksworld domain with 3 operators.

Table 3. Performance of Blackbox, STAN and our method on different problems

Problem	Blackbox	STAN	Our method	
Complete graphs	TT	TT	GT	TT
Sussman	0.02	0.028	0.005	0.006
Tw_rever4	0.03	0.03	0.011	0.021
Tw_rever5	0.06	0.03	0.023	0.04
Tower4	0.07	0.032	0.013	0.013
Tower5	0.21	0.07	0.024	0.025
Tower6	0.6	0.16	0.046	0.047
Tower9	111	10.53	0.225	0.225
T_largeA	0.82	0.53	0.079	0.08
T_largeB	4.34	2.63	0.289	0.3
T_largeC	—	82.143	1.792	1.8
T_largeD	—	—	4.508	4.52
Incomplete Graphs	TT	TT	GT	TT
Hanoi3d	0.11	0.039	0.019	0.176
Hanoi4d	1.41	0.061	0.035	1.81
Ferry	0.04	0.012	0.005	0.073
Monkeyt1	0.11	0.022	0.009	0.08
Monkeyt2	0.26	0.037	0.014	0.212

Table 4. Average and Standard Deviation of our method, STAN and Blackbox of solved problems

	Our method	STAN	Blackbox
Average	0.454	6.025	7.034
Standard Deviation	1.017	20.469	26.812

for solving planning problems as satisfiability problems [7]. Gerevini and Serina propose a new method for local search [5] in the context of the "planning through planning graph analysis" approach introduced by Blum and Furst [1]. In this latter work, the general search scheme is based on an iterative improvement process, which, starting from an initial subgraph of the planning graph, greedily improves the "quality" of the current plan according to some evaluation functions.

Our approach keeps some similarity with this latter work in the sense that it also uses an initial subgraph of the planning graph which is iteratively refined in three consecutive stages. The mechanisms used for the plan improvement are a combination of formal properties on graphs and heuristics to evaluate the best choice to restore unsolved preconditions. In this way, our approach exploits the advantages of a local search while keeping the theoretical completeness of systematic search techniques.

Local search techniques are incomplete in the sense that they cannot detect that a search problem has no solution [5]. However, one remarkable aspect of our approach is that it is able to detect unsolvable problems, while efficiently solving hard search problems.

7   Conclusions

In this paper we have presented a search method consisting in a four-process execution. Each process obtains a partial solution which is incrementally refined by the subsequent processes. The first stage simply obtains a graph from the problem which may comprise all the actions of a final solution. The basic graph selects the actions which form a part of a correct solution. The aim of the third stage is to guarantee completeness and optimality of the generated solution. And the POP is in charge of obtaining the final solution by finding a correct ordering among all the actions in the plan.

Our objective was to develop a new planning approach by taking advantage of the partial-order planning properties and reducing the inefficiency caused by the large search spaces generated by these planners. We have also shown that our method average outperforms other planning approaches as Graphplan or SATPLAN planners. This is a first prototype of the search method. The obtained results confirm that the POP is still a bottleneck mainly for those problems which give rise to an incomplete graph. For this reason we suggest that the introduction of the third stage will significantly reduce the amount of work done by the fourth stage.

Acknowledgments. This work has been partially funded by the Spanish Government CICYT-FEDER project 1FD7-0887.

References

1. Blum A. L., Furst M.L.: Fast Planning Through Planning Graph Analysis. Artificial Intelligence 90:281–300 (1997).
2. Chapman D.: Planning for Conjuntive Goals. Artificial Intelligence 32(3):333–377 (1987).
3. Fox M., Long D.: STAN public source code.
   http://www.dur.ac.uk/CompSci/research/stanstuff/ (1999)
4. Gerevini A., Schubert L.: Accelerating Partial-Order Planners: Some Techniques for Effective Search Control and Pruning. Journal of Artificial Intelligence Research 5:95–137 (1996)
5. Gerevini A., Serina I.: Fast Planning through Greedy Action Graphs. In Proceedings of the 16th National Conference on Artificial Intelligence (AAAI-99), 503–510 (1999)
6. Kautz H., Selman B.: Blackbox Planner 3.6.
   http://www.research.att.com/ kautz/blackbox/ (1999)

7. Kautz H., Selman B.: The role of domain-specific knowledge in the planning as satisfiability framework. In Proceedings of the 4th International Conference on AI Planning Systems (AIPS-98), 181–189 (1998)
8. Penberthy J.S., Weld, D.S.: UCPOP: A Sound, Complete, Partial Order Planner for ADL. In Proceedings of the 1992 International Conference on Principles of Knowledge Representation and Reasoning, 103–114 (1992). Morgan Kaufmann, Los Altos, CA
9. Sebastia L., Onaindia E., Marzal E.: Improving expressivity and efficiency in Partial-Order Causal Link Planners. In Proceedings of the 18th Workshop of the UK Planning and Scheduling Special Interest Group (PLANSIG-99), 124–136 (1999)
10. Sebastia L., Onaindia E., Marzal E.: A Graph-Based Approach for POCL Planning. In Proceedings of the 14th European Conference on Artificial Intelligence (ECAI-00), 531–535 (2000)
11. Onaindia E., Sebastia L., Marzal E.: 4SP: A four-stage planning process. In Proceedings of the ECAI-00 Workshop on New Results on Planning and Scheduling (PuK 2000), 115–129 (2000)

Map Drawing Based on a Resource-Constrained Search for a Navigation System

Hironori Hiraishi and Fumio Mizoguchi

Information Media Center
Science University of Tokyo
Noda, Chiba, 278-8510, Japan

Abstract. In this paper, we propose a map drawing method based on route-finding for a navigation system. To find a route, repeated searches expand branches from the current node to successive nodes until the goal node is reached. When the search space resulting from this process is drawn on a two-dimensional plane, we can regard it as a route map to a goal. We applied our method to a palm-top navigation system and developed a Resource-Constrained Heuristic Search (RCS) for making a route map. This allows us to set the memory and computing time limitations in advance. Palm-size PCs generally have insufficient memory to store entire maps on a practical scale and consequently must resort to downloading at the limited speeds typical of cellular phone communication. Thus, if we set a memory limitation, it can generate a map within the limited memory size; if we set a time-limit, it can achieve not only real-time route finding but also real-time map drawing. We implemented a palm-top navigation system that displays the current position on a map produced by the Global Positioning System (GPS) on a palm-size PC. Our experiment demonstrates that a route map generated by our RCS is sufficient for us to reach a goal along a designated route.

1   Introduction

Search is one of the most important AI methods and has been used for numerous problems, such as the eight puzzle and route-finding problems. It has often be used to find paths from a start state to a goal state. However, we propose that it can also be used for map drawing in a navigation system. To find a route, repeated searches expand branches from a current node to successive nodes until a goal node is reached. When the search space resulting from this process is drawn on a two-dimensional plane, we can regard it as a route map to a goal.

We applied our method of map drawing to a palm-top navigation system. Palm-size PCs generally have insufficient memory to store entire maps on a practical scale and consequently must resort to downloading at the limited speeds typical of cellular phone communication. To solve these problems, we developed a Resource-Constrained Heuristic Search (RCS) for making a route map, based on the ϵ-approximation search [Pohl,1973]. This allows us to set limitations for memory and computing time in advance, and it can generate the entire route

A. Nareyek (Ed.): Local Search for Planning and Scheduling, LNAI 2148, pp. 158–169, 2001.
© Springer-Verlag Berlin Heidelberg 2001

to a goal using the specified limitation. Thus, if we set a memory limitation, it can generate a map within the limited memory size; if we set a time limit, it can achieve not only real-time route finding but also real-time map drawing.

We implemented a palm-top navigation system that downloads map data from a server via cellular phone communication. Route finding is executed on the server side, and a route map is downloaded at the same time. We can also see the current position on a digital map produced by the Global Positioning System (GPS). Our experiment demonstrates that a route map generated by our RCS is sufficient for us to reach a goal along a designated route.

This paper is organized as follows. Section 2 explains map drawing based on a search and provides the algorithm of our RCS. Section 3 describes our palm-top navigation system. Section 4 concludes this paper and describes our future endeavors.

2   Map Drawing Based on a Search

A road map has a graph structure in which nodes represent crossroads and inter-nodal links represent roads. A search on the graph expands the branches from the current node to successive nodes. We can regard the search space as a route map to a goal.

Figure 1 shows an example of a route map constructed by such means. The search space always includes the route. Since it also contains successive nodes on the route, as shown in Fig. 1, it can construct road features like crossroads or side roads. We can traverse the route if we can identify which road we are presently on from GPS data.

We developed a Resource-Constrained Heuristic Search (RCS) to generate a route map using the search space. It is based on the ϵ-approximation search [Pohl,1973], and the evaluation function is $f(n) = g(n) + \epsilon \cdot h(n)$ [1]. In general,

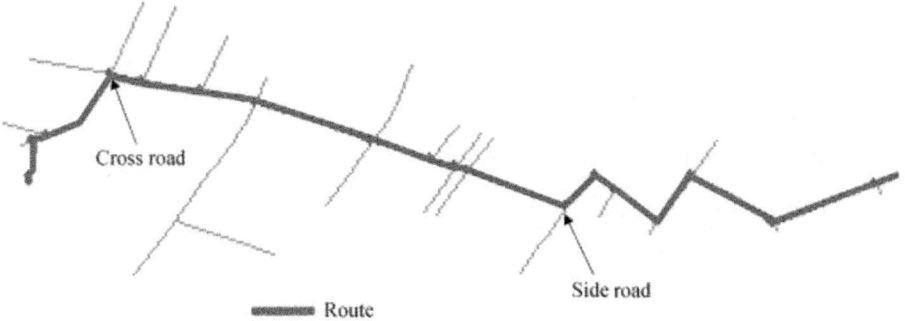

Fig. 1. Example of a route map constructed by a search space

---

[1] $g(n)$ is the actual cost from the start node to the node n, and $h(n)$ is the heuristic function that is the estimated cost from the node n to the goal node

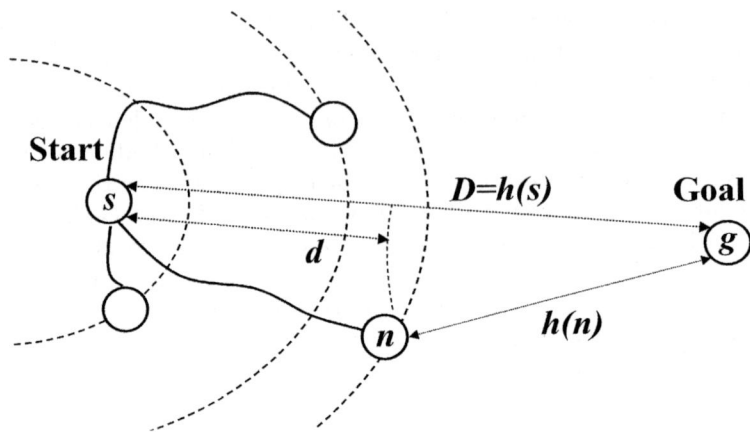

Fig. 2. Parameters for RCS

a larger ϵ value narrows the search space, reducing the resource to find a route to the goal.

Our method utilizes this characteristic and dynamically controls the ϵ value. Figure 2 shows the parameters for the RCS. D is the distance between the start node s and the goal node g and is also represented by $h(s)$. d is the effective distance that represents the progress of the search execution. It is calculated by $D - h(n)$, where the node n is the nearest node to the goal node. RCS controls the ϵ value as follows:

- $v - V < Lower$ then increase the value of ϵ
- $v - V > Upper$ then decrease the value of ϵ
- otherwise, no change

where $\epsilon \geq 1.0$ and the initial value of 1.0. v is the search velocity, which represents the progress of the search execution. It is calculated by $v = d/r$, where r is the amount of the consumption of the resource. V is the average velocity for finishing the search within the given resource and it is represented by $V = D/R$, where R is the limitation of the resource. Our method then controls the ϵ value to keep v equal to V.

The algorithm of our RCS is as follows.

Algorithm Resource_Constrained_Search($Start, Goal, R$)

```
1   OPEN ← {Start}, CLOSE ← φ
2   D = h(Start)
3   V = D/R
4   r = 0, d = 0, v = 0
5   ε = 1.0
6   min_h ← h(Start)
7   while OPEN ≠ φ
8       min_node ← get_min_node(OPEN, ε)
```

```
9        if min_node = Goal then
10           return route_to(Start, min_node)
11       CLOSE ← CLOSE ∪ {min_node}
12       OPEN ← OPEN ∪ expand_node(min_node, CLOSE, GOAL, ε)
13       min_h ← get_min_h(OPEN, min_h)
14       d = D − min_h
15       r = get_used_resource()
16       v = d/r
17       if v − V > Upper then
18          ε ← (ε − δ)
19       else if v − V < Lower then
20          ε ← (ε + δ)
21       if ε < 1.0 then
22          ε ← 1.0
23   return failure
```

We give the start node $Start$, the goal node $Goal$, and the limitation of the resource R. Lines 1 through 6 initialize each parameter. $OPEN$ is the set of end nodes of the found routes. The node that has minimum f in $OPEN$ is selected in the next loop. $CLOSE$ is the set of found nodes; its function is to avoid expanding a found node again. D is the distance that is the estimated cost from the start node to the goal node; it is initially set to $h(Start)$. V is the average velocity to finish the search within the given resource R. r, d, and v are set to 0. r is the amount of the resource consumption, d is the effective distance that indicates the progress of the search, and v is the effective velocity that is calculated by d/r. The initial ϵ is 1.0. The minimum h of all the found nodes is set at min_h.

The **while** loop in line 7 breaks if $OPEN$ is empty. This indicates that a route to the goal node does not exist, and thus this algorithm terminates and returns $failure$ (Line 23). If $OPEN$ is not empty, the algorithm repeats the following processes.

The **get_min_node** in line 8 selects the node that has the lowest f in $OPEN$. The selected node returns to the min_node. It has ϵ as its parameter. The evaluation function f must be recalculated with every change of the ϵ value. The heap tree cannot be allowed to accelerate in finding the minimum node in our algorithm, because the heap must be restructured when the ϵ value is changed. If the min_node is the same as $Goal$, the algorithm returns the path to the goal. The **route_to** in line 10 generates a path from $Start$ to the min_node.

The min_node is added to $CLOSE$ and is expanded by **expand_node** in line 12. The expanded nodes are stored in $OPEN$, and min_h is updated to the minimum h in $OPEN$. d, r, and v are recalculated in lines 14-16.

Lines 17 through 22 control the ϵ value. The ϵ value is determined by comparing v and V. δ is the parameter to increase or decrease the ϵ value. The ϵ value is reevaluated at every step in the algorithm.

If we give a time limit as a resource constraint, RCS becomes the Time-Constrained Heuristic Search (TCS) we proposed in [Hiraishi,1998]. TCS is an any-time algorithm [Dean,1988]. It can finish at exactly the time specified; a

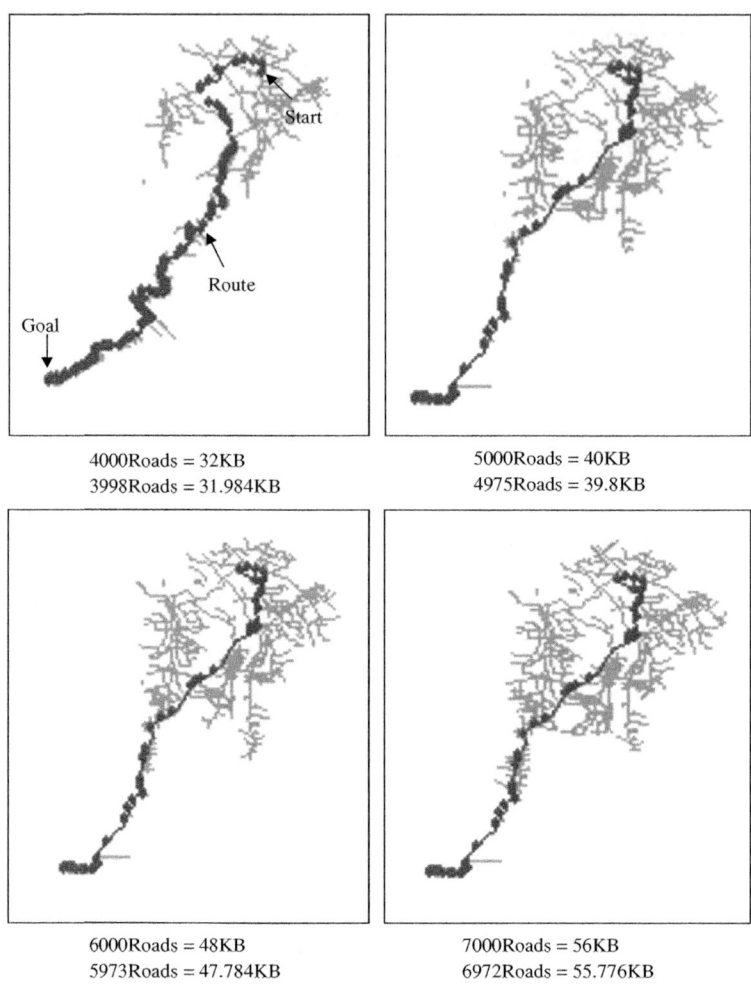

4000Roads = 32KB
3998Roads = 31.984KB

5000Roads = 40KB
4975Roads = 39.8KB

6000Roads = 48KB
5973Roads = 47.784KB

7000Roads = 56KB
6972Roads = 55.776KB

Fig. 3. Route maps by the RCS after setting a memory constraint. The upper description is the specified memory size (1 Road = 8 bytes), and the lower is the memory consumption for the route finding.

longer time limit yields a higher quality solution. These features are independent of machine performance and map structure [Hiraishi,1999].

Figure 3 shows route maps constructed by RCS after setting a memory constraint. The results are from route-finding between our university and Tokyo station[2]. The map area widened as the memory size was increased; we have demonstrated that RCS can construct a route map within the specified memory size.

---

[2] The length of the shortest route is 37.8km.

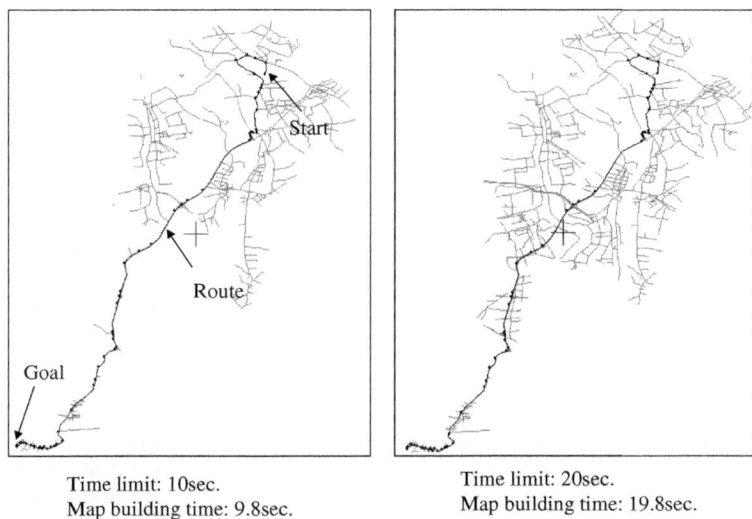

Time limit: 10sec.
Map building time: 9.8sec.

Time limit: 20sec.
Map building time: 19.8sec.

Fig. 4. Route maps by the RCS after setting a time constraint. The upper is the specified time and the lower is the actual time to finish the route finding.

Figure 4 shows route maps constructed by RCS after setting a time constraint. In this experiment, route finding was executed at the server (Sun Ultra-SPARC 168MHz WS), and the route map was drawn in the client (Celeron 366MHz PC) through cellular phone communication (communication speed: 32kbps). As soon as a node is expanded, the data of the node and link are sent to the client, and the node and link are then drawn in the client computer. As shown in Fig. 3, the map area widened as the time limit was increased, and RCS could finish constructing a route map in the specified time.

The time taken to draw a map includes the downloading time, and thus it is affected by the communication speed. However, the data is sent as soon as a node is expanded. RCS determines the ϵ value by calculating the search velocity at each step. The data transmission time is added to the node expansion time. Therefore, the communication speed is reflected in the search velocity. As a result, RCS can generate a route map within a specified time, independent of communication speed.

A heuristic search like A^* has advantages over an unformed search for map-making using a search space. For example, the search space of the Dijkstra algorithm spreads as a circle from the start node. In contrast, the search space of a heuristic search spreads toward a goal, so it can construct a route map faster using less memory. Figure 5 shows route maps constructed by other methods. The Dijkstra algorithm makes a map in the opposite direction to the goal. A^* is the heuristic search. However, since it always generates the optimal route, the route map is wider and A^* requires more map building time and memory consumption. RTA*[Korf,1990] is an any-time algorithm. It interrupts the search

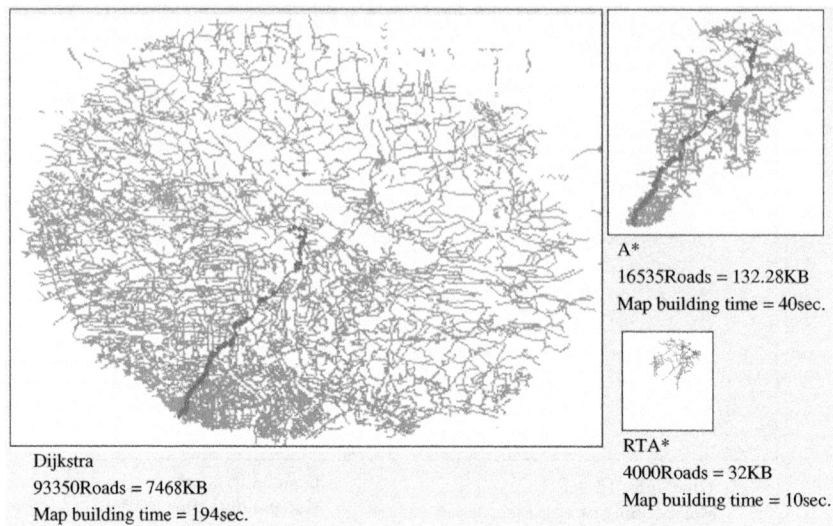

Dijkstra
93350Roads = 7468KB
Map building time = 194sec.

A*
16535Roads = 132.28KB
Map building time = 40sec.

RTA*
4000Roads = 32KB
Map building time = 10sec.

Fig. 5. Route maps by other search methods

at the time limit even if the entire route to a goal has not been found. Thus, if the search is interrupted before the goal node is found, a complete route map to a goal is not generated. The route map of the RTA* in Fig. 5 was interrupted after ten seconds. Therefore, the route map could not be generated, since RTA* could not find the goal node within ten seconds.

3   Palm-Top Navigation System

We implemented a palm-top navigation system (Palm-Navi) that has a map drawing function using our RCS. We used a CASIO palm-size PC E-503 (Fig. 6) running Windows CE 2.11[3]. It has a 6cm × 8cm color monitor with a resolution of 240 × 320 pixels. It also has an RC-232C port in which the GPS can be connected. We used a Sony GPS IPS-5000. GPS data comes at one-second intervals. A cellular phone was also connected in the RS-232C port. The maximum communication speed was 32Kbps.

3.1   User Interface

Figure 7 shows the output of our Palm-Navi. The Palm-Navi GUI is based on a digital map. The route map constructed by RCS is on the left. Our Palm-Navi was able to download a detailed map that is shown in the display area. The detailed map is in the center. The route is represented by red roads, and the

---

[3] CPU MIPS R4000 131MHz, 32MByte memory. The OS requires about 10MB of memory, so the available memory for applications is about 22MB.

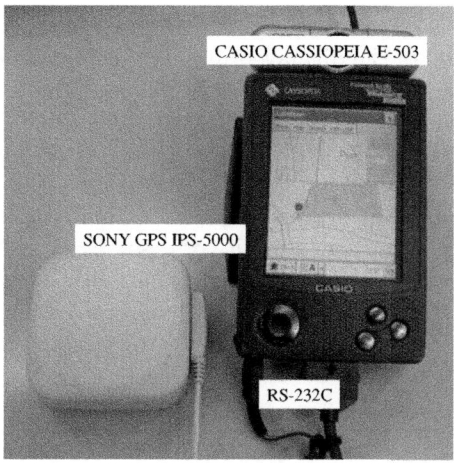

Fig. 6. Hardware for our palm-top navigation system

human shape indicates the current position and direction obtained from GPS data. When Palm-Navi acquires GPS data, the digital map scrolls automatically as the current position moves to the center of the display. Clicking on the map also activates the scroll feature, and the click point is displayed in the center.

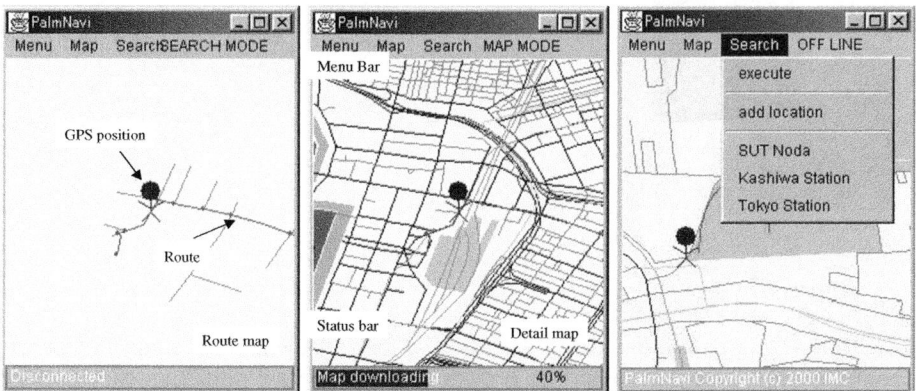

Fig. 7. User interfaces of our palm-top navigation system

There are four menus on the Menu Bar, and all Palm-Navi operations are performed by mouse clicks. In "Menu," we can save and load a previously downloaded map and set properties such as the server name. "Map" is used for zooming in and out. It also contains the menu to display the current position in the center of the display. We can execute route finding using "Search." When "execute" (right figure) is selected, the system finds a route from the current position to the center position of the displayed map. We can also register locations in the "Search" menu. The fourth menu is used to change the mode of our Palm-Navi. There are three modes: "MAP MODE," "SEARCH MODE," and "OFF LINE."

In "MAP MODE," Palm-Navi communicates with the server to download map data when the map is scrolled by acquiring GPS data and by clicking. Map data is also downloaded when we zoom in and out. In "SEARCH MODE," Palm-Navi does not communicate with the server even if the map is scrolled. This mode is used to refer to the overall route map. Though we must zoom out or scroll the map to utilize that feature, the map data is not downloaded in this mode. This allows us to look at the overall route map smoothly. "OFF LINE" is used without a server connection.

3.2   System Architecture

Figure 8 shows the system architecture for the Palm-Navi executing navigation and MapServer sending map data. The digital map is stored in the MapServer.

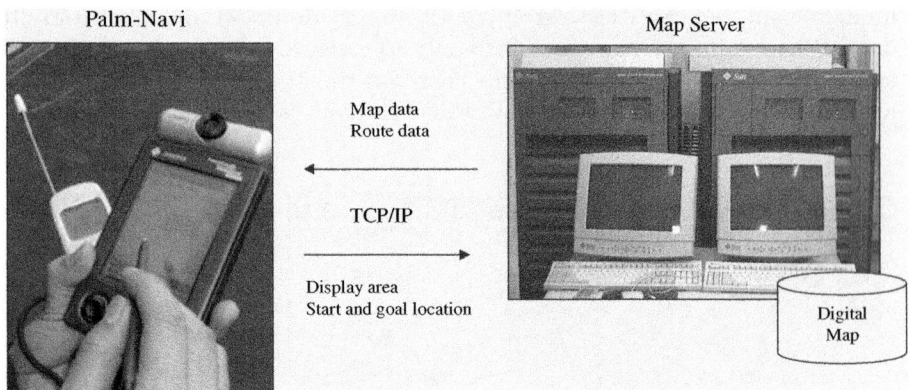

Fig. 8. System architecture

Palm-Navi communicates with MapServer to download a map and execute route finding. The data of the display area is sent from Palm-Navi to MapServer as a request for map downloading. The data form is as follows:

$$< Level, Longitude, Latitude, Width, Height >$$

Level indicates the scale of the map, *Longitude* and *Latitude* are the geographical coordinates of the lower-left corner of the display area, and *Width* and *Height* are the longitude width and latitude height. Thus the display area represented by geographical coordinates is sent to the MapServer.

MapServer returns only the map data that can be contained in the display area. MapServer temporarily stores the map data already sent to avoid re-sending the same data. Map data from MapServer is shown in Table 1. Map data consists of three types: road data, geographical data representing geographical features and buildings, and string data. *DataID* in each data format labels the data as road data, geographical data, or string data. *ColorID* specifies the color in which data is to be displayed, and Level indicates the map scale.

Table 1. Map data format

	Data Format
Road data	$< DataID, ColorCode, Level,$ $Num, Longitude, Latitude, ... >$
Geographical data	$< DataID, ColorCode, Level,$ $Num, Longitude, Latitude, ... >$
String data	$< DataID, ColorCode, Level,$ $Longitude, Latitude, String >$

Num in the road and geographical data indicates the number of the pair of *Longitude* and *Latitude* following *Num*. A road is drawn on the map by connecting successive links between a start point and an end point; geographical features and buildings are also presented on the map. *Longitude* and *Latitude* in string data are the geographical coordinates to position *String*. The geographical coordinates of a start position and a goal position are sent to MapServer in the following form for a route-finding request:

$$< Slongitude, Slatitude, Glongitude, Glatitude >$$

Slongitude and *Slatitude* define the start point, and *Glongitude* and *Glatitude* specify the goal coordinates. The map data for the route map is returned to Palm-Navi during the route finding. The data is in the same format as the road data, but the *DataID* is different. When the route finding is completed, the route data is sent to the Palm-Navi. The data for the route also takes the same format, but the *DataID* is different.

3.3   Use of Palm-Navi

In this section, we provide some evaluations of the actual use of our Palm-Navi. Several examples of our use of the Palm-Navi are shown in Fig. 9.

1. Palm-Navi connects to the server using a dialup connection (Picture 1) to download map data. The dialup connection requires about 30 seconds, and the downloading and drawing of the detailed map in the display area take about 20 seconds.
2. Palm-Navi is installed in a car after obtaining a route map by executing route finding (Picture 2). We use suction cups to attach our Palm-Navi so that we can install it anywhere on the window. Picture 2 depicts connecting the power supply and GPS to the Palm-Navi.

3. Use while driving is illustrated in Picture 3. We are driving along the route while looking at the route map and checking our current position and direction, as shown in Picture 4.

There is some concern that Palm-Navi may distract the driver because it is difficult to view a map on its small display. However, the route map displayed on the Palm-Navi is so simple (see Picture 4) that we can see and easily understand the route and current position at a glance, even on such a small display. Palm-Navi thus aids, rather than distracts, the driver. In our navigation test, we chose a route in the Tokyo area that we had never driven before, and we arrived at the proper destinations without getting lost. The simple route map contained enough road features for us to navigate along the route. Thus, the route map generated by our RCS is sufficient for us to reach a goal along a designated route.

4   Conclusions

In this paper, we proposed a map drawing method based on route finding for navigation systems. The search was used only for path finding to a goal state; we used a map drawing method that regards the search space as the route map to a goal.

We applied our method to a palm-top navigation system and developed a Resource-Constrained Heuristic Search (RCS) to generate a route map. This method allows us to set memory and time limitations in advance, and therefore it is suitable for a palm-size PC with little memory and a small display. Although

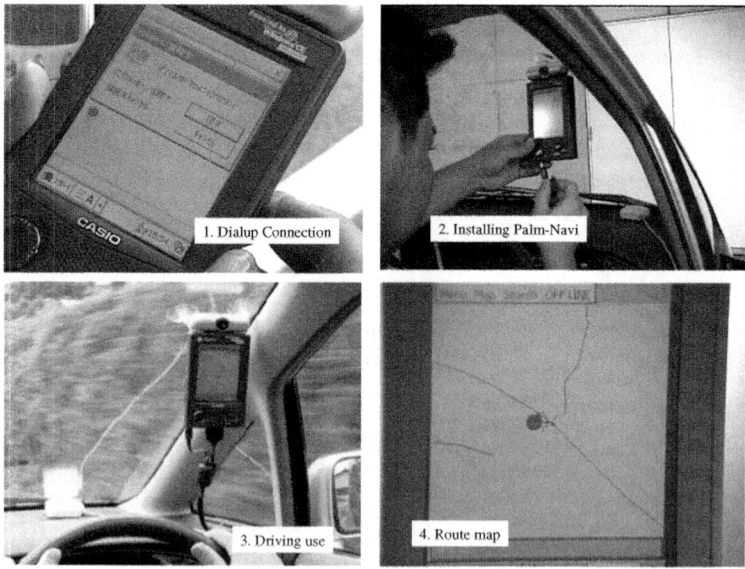

Fig. 9. Using the Palm-Navi

the route map constructed by the search space is simple, it provides the minimum required information, including road features such as crossroads and side roads. We clarified that we can navigate along an actual route using our system.

Our digital map separates the data for map drawing from road data for route finding. Therefore, the route map consists solely of road data. A future project is to design a map database to receive and store road data, geographical data, and string data during route finding.

References

[Dean,1988] Thomas Dean and Mark Boddy, "An Analysis of Time-Dependent Planning," In Proc. of the Seventh National Conference on Artificial Intelligence, pp. 49-54, 1988.

[Hiraishi,1998] Hironori Hiraishi, Hayato Ohwada, Fumio Mizoguchi, "Time-Constrained Heuristic Search for Practical Route Finding," Lecture Notes in Artificial Intelligence 1531, pp. 389-398, 1998.

[Hiraishi,1999] Hironori Hiraishi, Hayato Ohwada, Fumio Mizoguchi, "Intercommunicating Car Navigation System with Dynamic Route Finding," In Proc. of the IEEE/IEEJ/JSAI International Conference on Intelligent Transportation Systems (ITSC'99), pp. 248-289, 1999.

[Korf,1990] Richard E. Korf, "Real-Time Heuristic Search," *Artificial Intelligence* 42, pp. 189-211, 1990.

[Pohl,1973] Pohl, I., "The avoidance of (relative) catastrophe, heuristic competence, genuine dynamic weighting and computational issues in heuristic problem solving," In Proc. of the Third International Joint Conference on Artificial Intelligence, pp. 20-23, 1973.

Author Index

Lecture Notes in Artificial Intelligence (LNAI)

Lecture Notes in Computer Science

GPSR Compliance

The European Union's (EU) General Product Safety Regulation (GPSR)
is a set of rules that requires consumer products to be safe and our
obligations to ensure this.

If you have any concerns about our products, you can contact us on
ProductSafety@springernature.com

In case Publisher is established outside the EU, the EU authorized
representative is:

Springer Nature Customer Service Center GmbH
Europaplatz 3
69115 Heidelberg, Germany

Batch number: 09624486

Printed by Printforce, the Netherlands